THE PRACTICAL GUIDE TO ENVIRONMENTAL COMMUNITY RELATIONS

THE PRACTICAL GUIDE TO ENVIRONMENTAL COMMUNITY RELATIONS

Carol J. Forrest
Renee Hix Mays
Equinox Environmental Consultants, Ltd.

John Wiley & Sons, Inc.

New York ■ Chichester ■ Brisbane ■ Toronto ■ Singapore ■ Weinheim

Library of Congress Cataloging-in-Publication Data

Forrest, Carol J., 1959–
    The practical guide to environmental community relations / Carol
J. Forrest & Renee Hix Mays.
      p. cm.
    Includes bibliographical references and index.
    ISBN 0-471-16388-0 (cloth : alk. paper)
    1. Industrial management—Environmental aspects. 2. Social
responsibility of business. 3. Environmental policy—Citizen
participation. 4. Public relations—Corporations. 5. Issues
management. I. Title.
HD30.255.F67 1997
658.4'08—dc20   96–41858

Printed in the United States of America

10 9 8 7 6 5 4 3 2 1

ACKNOWLEDGEMENTS

Our thanks to the following people for sharing their insights and experience into community relations:

Carole Braverman, U.S. Environmental Protection Agency, Region V; Claire Brechter, Claire Brechter Public Relations; Roberta L. Harper, Unicycle Communications; David Hastings, JBF Associates, Inc.; Penny T. Hill, Communication Strategies; Greg R. Michaud, Illinois Environmental Protection Agency; Michael D. Moss, Chemical Industry Council of Illinois; Cindy Newman, The Dow Chemical Company; John P. Perrecone, U.S. Environmental Protection Agency, Region V; Joseph Pokorny, Corporate Public Affairs Consultant; Stephanie Reith, Rust Environment & Infrastructure; Jack Toslosky, Chemical Industry Council of Illinois; Deborah Volkmer, Roy F. Weston, Inc.; Mary Wenska, Black & Veach Special Projects Corp; Paul H. Wyche, Jr., Safety-Kleen Corp.; and Suzanne Zoda, Environmental Communication Solutions Inc.

Special thanks to Philip W. Cotter, for preparing the graphics for this book.

Thanks, as well, to Corinne Bodeman, Thomas Enno, Kathleen J. Getty, Ginger Griffin, Kenneth M. Jurish, Esq., and Kathleen Madigan for their encouragement and support.

CONTENTS

CHAPTER

ADDRESSING COMMUNITY CONCERNS ABOUT ENVIRONMENTAL ISSUES— THE IMPORTANCE OF ENVIRONMENTAL COMMUNITY RELATIONS

I know of no safe repository of the ultimate powers of society but the people themselves; and if we think them not enlightened enough to exercise their control with a wholesome discretion, the answer is not to take it from them, but to inform their discretion.

—Thomas Jefferson

THE MANY CIRCUMSTANCES OF ENVIRONMENTAL CONCERNS

- Cleanup of a contaminated site is halted as community residents launch a legal challenge against the method that those responsible for the site have selected for remediation.
- Parents and teachers at a public school call for an investigation to determine if substances in the building are making students and staff sick. Midway into the investigation, they also become concerned as to whether the right tests are being performed to detect contamination that may be present, and how they will know what the test results mean.

1

- Plans to move a manufacturing operation into an industrial park are stymied after nearby residential neighbors, who fear that there may be odors associated with the company's processes, stage a protest at the town's planning commission meeting. Alarmed at newspaper coverage of the incident, the company abandons its plans to relocate to that community—forfeiting $100,000 already spent on the selection of the site.
- Industrial neighbors of a facility oppose its air permit application out of concern for the effects the facility's emissions might have on their own workers.

Virtually no business enterprise or government entity is immune from the potential for public interest in or concern about environmental issues. Public concern over environmental issues may arise as a result of any number of different situations, including:

- Permitting activities under the Resource Conservation and Recovery Act (RCRA), the Clean Air Act, and the Clean Water Act.
- Investigations and/or cleanups of known or suspected contaminated sites under the Comprehensive Environmental Response, Compensation, and Liability Act (CERCLA—also known as Superfund) or other programs.
- Spills or accidents involving hazardous chemicals, wastes, or petroleum products.
- Publicity about emissions from, or chemical storage or use at, a facility.
- Environmental effects of certain products or practices.
- Development of land or coastal areas and exploitation or harvesting of natural resources.
- Environmental issues at *other* sites or facilities that create concerns about similar scenarios elsewhere.
- Side issues to disputes such as those that occur over land use planning or within the rhetoric surrounding labor conflicts.

Within the public arena, environmental issues are rarely viewed only in terms of their scientific, technical, or regulatory merits. Instead, they are invariably intertwined with other issues and perceptions regarding health and safety, property values, social justice, and overall quality of life. As such, environmental issues have the potential to create significant controversy, concern, and opposition among community residents.

Failure to deal effectively with the questions and concerns that can arise in conjunction with environmental issues can lead to serious consequences for both public and private sector organizations. The emotional cost to community residents can also be extremely high. In many cases, however, much, if not all, of the anger and fear that community residents feel can be relieved through a substantive dialogue that either corrects misinformation or provides a platform for discussing actions that could, should, or are taking place to remedy a situation. The process used to work with community residents and other stakeholders to resolve their concerns and answer their questions is the subject of this book.

WHAT IS COMMUNITY RELATIONS?

There are a number of terms that refer to the process of working with the public in regard to environmental issues. "Community relations," "community outreach," "public participation," "public involvement," and "stakeholder involvement" are all terms used by the U.S. Environmental Protection Agency (U.S. EPA) and other government agencies as well as by many private sector practitioners. These terms are often used interchangeably. The authors have chosen to use the term "community relations" in this book because it is probably the most commonly recognized of the five terms for the broad range of activities and scenarios that we discuss. Additionally, we believe that the term "community relations" connotes an ongoing and enduring process that tracks with many of the concepts we present.

Community relations is generally defined as two-way communication to enhance public understanding of environmental issues and to encourage input from the public so that their concerns are considered in organizational decision-making processes. Although communication is a major component of community relations, in practice, community relations is infinitely more than the simple release of information.

True community relations is an *interactive process* involving both substantive, two-way communication and the building of positive relationships with key stakeholders. Each reinforces the other. Open communication enhances and deepens relationships and builds trust and credibility. Good relationships create an atmosphere conducive to the discussion of serious topics, such as the risk or perceived risk to human health and the environment posed by a facility, site, or project. In the words of community relations veteran John P. Perrecone, community involvement coordinator for U.S. EPA Region

V's Office of Public Affairs, "Trust and credibility are the most important components in communicating about environmental issues. It takes time to build these credentials—that time must be spent in the community learning their concerns and issues."

As we will discuss later in this book, the level of trust and credibility with which stakeholders view a messenger has a direct influence on how willing they are to accept his or her information or to work toward resolving the issue in question.

The community relations process also deals with the concept of stakeholder empowerment and control. As we illustrated at the beginning of this book, public concerns can seriously hinder or block the operation of a facility, the investigation or cleanup of a contaminated site, or the implementation of a project or development. The community relations process provides a framework for mediating between stakeholders and the organization responsible for the facility, site, or project to achieve resolutions that either are acceptable to all parties or minimize the burden that any one party has to bear.

The concepts of two-way communication and empowerment are inexorably linked. Communicating with stakeholders about environmental issues invariably invests in them some degree of control. In cases in which communication is used to correct simple misunderstandings, lack of information, or mild differences of opinion, the community relations dialogue has no noticeable effect on the operation of a facility, site, or project. Thus, the empowerment dimension of community relations is frequently overlooked. In most of these cases, stakeholders have received sufficient information about an issue so that the majority either feel comfortable with the decisions that the organization's managers have made, or their questions or concerns have been answered well enough so that they are not in active opposition.

In some other cases, however, public comments or concerns lead organization managers to modify their operations or decisions in some way to accommodate stakeholder wishes or demands. In still other cases, public input involves an actual seat at the decision-making table. Stakeholders are empowered to veto, accept, or modify the way a facility or project operates or a site is cleaned up. In such situations, a good dialogue is essential to provide stakeholders with enough information about the scientific, technical, and regulatory aspects of the issue so that good decisions can be made. Figure 1.1 illustrates the hierarchy of communication within the context of the community relations process.

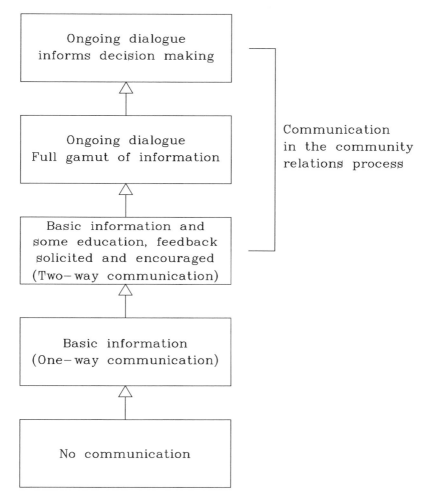

Ongoing dialogue
informs decision making

Ongoing dialogue
Full gamut of information

Basic information and
some education, feedback
solicited and encouraged
(Two–way communication)

Basic information
(One–way communication)

No communication

Communication
in the community
relations process

■ **FIGURE 1.1 Communication Hierarchy in Community Relations**

Effective community relations may involve a complex, long-term program with numerous formal and informal meetings, workshops, and open
houses and much written information, or it may involve as little as an occasional informal chat with a neighbor who has questions or wishes to make
comments about an environmental issue regarding a facility, site, or project.
The scope of a given community relations effort will depend on a number of
variables that we will explore in this book. Among these variables are the
level of concern that stakeholders have about a particular issue (which may

or may not be tied to the actual risk posed by the situation in question), the manner in which stakeholders wish to engage in a dialogue, and the duration over which concerns or questions may persist.

For example, questions or concerns regarding the cleanup of minor contamination that poses little risk to human health or the environment may last only until the cleanup is completed. Questions or concerns regarding the operation of a manufacturing facility may persist as long as the facility is in existence. Such a facility may need to engage in an ongoing dialogue with its neighbors and other key stakeholders to ensure that they are comfortable with its operations and its environmental performance. Regardless of the magnitude and "formality" of the community relations effort, however, the approach will be the same and will track with the concepts and use the techniques that we describe in this book.

Myths About Communicating With the Public

Managers of facilities, sites, or projects that are subjects of community interest or concerns are often reluctant to engage in community relations, either because they believe that there is no real risk to the community, and, thus, no reason to respond to residents' concerns, or because they believe that providing information about the issue will create or worsen community concerns.

Even in cases in which concerns on the part of community residents are based on misinformation, it is important to remember that, when dealing with the public, perception equals reality. This is not to say that perceptions cannot be changed—they can, and that is often an important goal of a community relations program; however, if community residents perceive risk, then these perceptions need to be addressed. Moreover, failure to acknowledge such concerns, even if they are in error, can fuel public distrust or anger that itself can become a much greater problem than the original issue.

Properly done, community relations *does not* heighten concern or create concern where it did not previously exist. If an issue is one that is already public or likely to become public in the future, then there is little to be gained by not engaging in a dialogue with stakeholders. Communication should take place early and often to build familiarity with the issue and convey a sense that those involved in or responsible for the issue are committed to an open dialogue.

Anger at being "kept in the dark" frequently augments or eclipses other concerns that community residents might have. It is important to note that, to the public, omission of information is often considered a form of lying and can evoke a significant backlash in its own right. In the authors' experience, the public is typically more accepting of a certain level of risk than of substantive omissions of information or the belief that they have been lied to. Community residents are often as much influenced, reassured, or dismayed by the way managers of a facility, site, or project listen (or fail to listen) to their concerns and respond (or fail to respond) to their questions or suggestions as they are by the technical or regulatory merits of the issue in question.

Determining Specific Community Behavior

Environmental issues don't exist in a vacuum. An important aspect of the community relations process is determining how stakeholders will respond to environmental issues or, if an issue is already known, what other variables within the community need to be considered to engage stakeholders in an effective dialogue. For example, will an application to increase emissions to accommodate the expansion of a manufacturing facility be viewed as a nonissue—the necessary price of more jobs, or nothing of concern in a community where air and water quality are good? Or will it be viewed as an intolerable danger to the health of community residents? Either scenario is possible, depending on a number of variables, including:

- Existing attitudes toward the facility and its role in the community.
- The existence of other environmental issues in the community that may have sensitized residents to the risk (or perceived risk) of industrial activities.
- Other community issues, such as political battles between stakeholder groups, economic growth or economic decline, disagreements about land use, or concerns about property values.

One of the keys to successfully addressing community concerns about environmental issues is understanding the *context* in which a specific issue is viewed. Chapter 2 of this book explores how managers responsible for facilities, sites, or projects can examine contextual variables to determine the likelihood that stakeholders will have concerns about an environmental issue, and what the nature of these concerns might be.

The Link Between Community Relations and Environmental Performance

Many concerns about environmental issues are based on misinformation, lack of information, or differences of opinion. The emphasis of the community relations effort in such cases will be on communication. Other concerns may result from accidents or the discovery of contamination or other unintended consequences that may pose a risk to human health and the environment. In these cases, communication is important; however, actions to remedy the problems must also be taken.

Community relations is not a panacea. It is not intended to smooth over with talk issues that clearly require action on the part of the organization responsible for a facility, site, or project. In order for positive relations with the community to be sustainable, organizations must demonstrate an ongoing commitment to protecting human health and the environment by establishing and maintaining a high level of environmental performance.

What Drives Conflict About Environmental Issues?

How do conflicts over environmental issues arise? American culture includes a number of deeply ingrained core values that can come into play over environmental issues. These include:

- The right to do what one wishes with one's private property.
- The free-market system.
- The rights of the individual (or business enterprise) to do as he or she chooses.
- Freedom of speech.
- Representation in the formation of the laws that govern us.
- Local control of local issues.
- The right to know about toxic chemicals and the environmental performance of companies in one's community.
- The right to live in a clean environment and be protected against adverse health effects.

Some of these core values are expressly articulated in the U.S. Constitution. Others—such as the right to know about toxic chemicals and the right to live in a clean environment—are only recently articulated beliefs.

Conflict over environmental issues frequently involves controversy over which of these core values are preeminent—or, often, how the issue in question is defined in regard to these core values.

For example, community residents concerned about emissions from a facility will place their right to live in a clean environment ahead of a facility's right to emit. The facility's managers may say that they aren't violating that right—that their emissions are below the levels allowed by regulation, and don't pose an undue risk. This being the case, facility managers emphasize *their* right to use their property (the facility) as they see fit, and their right to run their business to participate in the marketplace unfettered by undue constraint.

Such cases are fraught with complications. Are the emissions indeed posing an unacceptable risk? By whose standards? What is the role of perception? Of lack of information or misinformation? Do the neighbors perceive that, in addition to risk to their health, the value of their own private property is being degraded as well? Who was "here" first?—This is a question that is often asked, but never leads to a satisfactory conclusion, since everyone is "here" now, and everyone has a vested interest. Whose "rights" should prevail? Is there a way to resolve a conflict such as this so that, at best, everyone is satisfied, or, at least, no one party has to shoulder a disproportionate share of the burden?

The role of community relations frequently involves attempting to determine how—or whether, in a given situation—these core values can be reconciled when they appear to conflict and working with stakeholders to achieve acceptable resolutions. Thus, community relations is not simply about environmental issues or about communication—it's about the fundamental hopes, fears, and beliefs that all of us hold.

THE INEVITABILITY OF DEALING WITH THE PUBLIC

There are a number of reasons why the need for good relations with community residents and other stakeholders is increasing. The most tangible and obvious of these reasons has to do with increased regulatory requirements for involving the public in environmental decision making and for disclosing information on environmental performance and conditions. These regulations, which are discussed in detail in chapter 7, have driven the need for a positive dialogue with the public. A short description of some of the major environmental laws appears at the end of this chapter. Additionally, certain social trends that have sensitized the public to environmental issues and increased the overall level of community-based activism are also driving the need to work with community stakeholders.

Regulatory Requirements

Over the past 15 years, a number of regulations have been promulgated that provide community residents access to the decision-making processes regarding permitting and the cleanup of certain contaminated sites or that require the reporting or public disclosure of information about emissions, chemical storage, and other activities at facilities.

Requirements for community relations/public involvement are especially prominent in CERCLA (Superfund) and RCRA. Requirements for community relations/public involvement are also included in regulations promulgated under the Clean Water Act. In the case of CERCLA, for example, the regulations allow for public comment on the remedial alternatives being considered for the cleanup of Superfund sites. Guidances under CERCLA also require a number of other community relations activities, such as public notices, public hearings, and establishment of information repositories, to keep community residents and other interested parties informed about activities at CERCLA sites.

The regulations under RCRA and the Clean Water Act allow for public comment during the permitting process. Additional regulations, adopted in December 1995, provide for expanded public participation under RCRA, which regulates the generation, transportation, treatment, storage, and disposal of hazardous wastes. These regulations require informal meetings before permit applications are submitted to the state or federal environmental agency, the posting of signs on facility property and placement of display advertisements in local newspapers providing information on the upcoming permitting activities, the issuance of fact sheets, and other activities designed to inform residents about permitting activities and solicit their input.

Reporting of environmental information, such as emission levels and storage quantities of certain hazardous chemicals, is required under the Emergency Planning and Community Right-to-Know Act (EPCRA), Title III of the Superfund Amendments and Reauthorization Act (SARA) of 1986. The best known of the data available to the public under EPCRA is the Toxics Release Inventory (TRI), which lists the amounts of certain toxic chemicals emitted to the land, water, and air or injected into the ground by manufacturing facilities. Although reporting under EPCRA does not require public involvement in decision making—there are no public hearings or public comment periods—the public availability of this information means that the reporting facilities may be asked by community residents and environmental activists to explain why they emit certain chemicals. These inquiries can result in the need for a substantial dialogue.

Similarly, recent requirements for certain facilities to disclose, by mid-1999, "worst-case scenario" and spill history information under Section 112(r) of the Clean Air Act are also likely to result in the need for a substantial dialogue with facility communities. This regulation requires facilities that store "threshold quantities" of certain chemicals to prepare "risk management plans," which must provide information on the extent of potential off-site consequences in the event of an accidental release of the chemicals in question. This type of information, which will be made available to the public, will invariably raise some eyebrows—and some questions—among residents who live near these facilities.

Regulations have a ripple effect. The increase in questions and concerns voiced by the public after the TRI data were first released in the late 1980s is a case in point. The TRI data have provided information that has both sensitized residents to emissions and chemical use in their communities and provided a wealth of information for activist organizations to use to question facility managers about their operations and environmental performance. Industry leaders, the U.S. EPA, and activist groups alike credit the TRI with providing the impetus for businesses to begin to address the environmental concerns expressed by the public.

Additionally, the public availability of information under programs such as CERCLA and RCRA has prompted environmental activists and community residents alike to demand information about *other* facilities, sites, or projects, regardless of whether they fall under regulations requiring disclosure or public involvement. For example, community residents who have obtained information about a local Superfund site through the regulatory process may believe that they should be able to obtain the same sort of information—and have an opportunity to comment on remedial alternatives—about a site undergoing a voluntary cleanup. As a result, many businesses and government entities find that they need to initiate "voluntary" community relations programs to meet the expectations of community residents and other stakeholders. These organizations have discovered that working with the public and considering public concerns and wishes during the decision-making process can mean the difference between serious opposition or public acceptance of the issue in question.

Increasing Use of Risk-Based Approach to Site Cleanups   With the increasing acceptance and use of risk-based approaches to cleaning up contaminated sites, the question of health effects will no doubt become quite prominent in communities in which such cleanups are performed. The use of risk assessment to guide the cleanup of sites, rather than attempting to clean all of them up to "pristine" levels, can easily raise questions and concerns among neighbors about risk. Organizations responsible for such cleanups will need to work with site neighbors to educate them about how exposure to any remaining contaminants will be minimized or prevented at a particular site to protect neighbors' health.

Increasing Local Control of Environmental Regulation Another important trend in the regulatory arena that is likely to lead to increased involvement at the community level is the gradual shift in responsibility for regulation and enforcement from federal to state and local jurisdictions. As responsibility for protecting human health and the environment moves to state and local government jurisdictions, area residents and officials will necessarily be looking more closely at how local facilities, sites, or projects are managed. The authors of this book, and others who specialize in community relations, believe that this trend will mean that managers responsible for facilities, sites, or projects in many communities will have to increase their communication with local stakeholders.

Social Trends Driving the Need for Community Relations

A number of social trends are influencing the way the public views both environmental issues and the role that community residents and other stakeholders should play in addressing these issues. These trends, coupled with increased availability of information and access to decision making through regulatory programs, have created conditions that are ripe for the expression of environmental concerns on the part of the public.

The "Overinformed, Undercomprehending Public" and the "Mainstreaming of Activism" In addition to the information that is available under the major environmental regulations, the public is bombarded daily with other information pertaining to environmental and health issues. This information includes news and feature stories on a broad range of environmental and health topics. Some of these stories are highly speculative and most provide only very basic information, with little explanation of science or technical processes, or, in the case of stories about fines or violations, of the regulations. Actual links, suspected links, and perceived links among environmental conditions, chemical exposures, and health effects have heightened health concerns among the public. Since people perceive that their own health, or the health of their children or loved ones, may be affected, the level of concern is understandable. Publicity about the discovery of contamination, such as at Love Canal, New York, and Times Beach, Missouri, and accidents and spills, such as the methyl isocyanate release in Bhopal, India, or the Exxon Valdez oil spill in Alaska, have heightened community concerns about the risks posed by chemicals and industrial processes.

Although the public receives many messages about environmental and health issues, it receives little education regarding what these messages actually mean, especially within the context of specific issues that may be unfolding in specific communities. Thus, the public is "overinformed, but undercomprehending" when it comes to environmental issues.

At the same time, the concept of community activism has gone mainstream. No longer are people who question the environmental impact or potential risk posed by a site or facility considered radical. Overall, Americans believe that they have a right to know about corporate and government environmental performance and other issues that may affect their health. Numerous high-profile news stories over the past 15 years have shown activists and "average" citizens protesting, often successfully, against situations involving actual or perceived environmental threats. The activities of these individuals and groups have served both to legitimize community activism and to provide a template for other members of the public who wish to oppose situations that concern them—what we call the "mainstreaming of activism."

Although a number of news stories over the past few years have reported that many national environmental groups have experienced decreases in their memberships, these stories have also stated that local "grassroots" groups formed to address local environmental and community issues are on the increase. Casual observations by the authors and others who work in the community relations field suggest that local groups are indeed on the increase.

The mainstreaming of activism has much broader implications than increased membership in local environmental or community groups, however, since persons who aren't "activists" in the usual sense of the word may also express concerns about environmental issues or oppose operations at a facility, site, or project. As many community relations specialists can attest, these "average" citizens may also follow the activist template of vocal opposition when they encounter an issue that personally affects them (or that they perceive as personally affecting them). Serious issues *can* awaken the "sleeping giant" of the average citizenry. As Figures 1.2*a* and 1.2*b* illustrate,

No Major Environmental Issues In Community

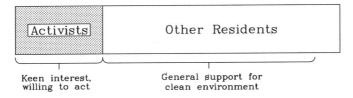

■ **FIGURE 1.2a Specific, Expressed Concerns About the
Environment Among Nonactivist Residents May Be Dormant in
Communities that Aren't Facing Major Environmental Issues**

Expansion of facility that handles
hazardous chemicals

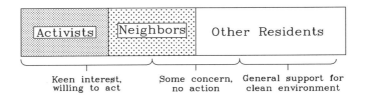

Local aquifer used for water supply
endangered by contamination from
abandoned site

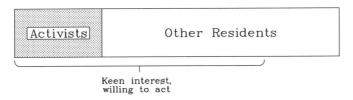

■ **FIGURE 1.2b Some Local Environmental Issues Can Lead
to Substantial Community Concern**

the segments of a population that may mobilize to actively oppose or speak out on an issue will vary according to the issue—and the active voicing of concerns isn't confined only to activists.

The new "activism" involves far more than environmental issues. Casual review of any major metropolitan newspaper over the period of a week or so will reveal stories about a variety of issues under active protest by "average" community residents. The subject of protest or concern may be a proposed multiscreen movie theater, a jogging path or basketball court, proposed low-income housing, or a shopping mall; however, the tactics and many of the basic types of concerns are the same as if the subject was a landfill or an incinerator. Thus, the activist template has become a part of the cultural landscape within which managers of facilities, sites, or projects must work.

The Redefinition of "Community" Another interesting concept that is gaining currency among some environmental and community activists and that the authors have themselves encountered in several communities involves a redefinition of the term "community." Just as residents involve themselves in the operation of schools, parks, and other "common elements" of their community, residents of some communities are also beginning to view manufacturing facilities or government facilities, such as military installations, within or near their communities' borders, as "common elements." Residents in these communities are requesting or demanding information on facility operations and environmental performance or on site cleanups, along with a say in such issues as chemical use and pollution prevention initiatives.

It is likely that, as information on the potential for off-site consequences from the accidental release of chemicals (under Section 112(r) of the Clean Air Act) is made available to the public, this trend will accelerate, and community activists and leaders will assert that they should have input into the decision-making processes of facilities that they believe could affect their lives.

Environmental Justice Environmental justice is a relatively new concept that speaks to concerns that minority and low-income communities have been disproportionately affected by pollution and degraded environments. A number of studies by different organizations have been performed correlating demographic data and the presence of RCRA permitted facilities and/or contaminated sites. Some of these studies appear to show that minority or low-income populations have been disproportionately affected by pollution or in the siting of RCRA facilities; however, other studies have failed to show pervasive patterns. The debate continues to rage.

Regardless of the conflicting studies, environmental justice is an extremely important issue for what it tells us about the need to proactively reach out and include *all* potential stakeholders in the dialogue and the decision-making processes that surround environmental issues. It is *not* acceptable to exclude—or assume a lack of interest on the part of—potential stakeholders, and a number of minority and social leaders are urging neighborhood residents to become more involved in environmental issues in their communities to ensure that they have input into decision-making processes. We explore environmental justice more fully in chapter 8 of this book.

THE ROLE OF ORGANIZATIONAL CULTURE IN COMMUNITY RELATIONS

Environmental community relations is a relatively new discipline. Although some organizations have embraced its concepts and are committed to working with affected stakeholders, community relations is still a foreign concept to many other organizations. The ease with which community relations concepts are adopted and implemented by an organization has a lot to do with its organizational culture. In many cases, new ways of thinking about community residents and their right to engage in a substantive dialogue with an organization's managers need to be internalized before the community relations process can be effectively employed.

All organizations, regardless of whether they are for-profit businesses, not-for-profit entities, or federal, state, or local units of government, have their own sets of beliefs and modes of behavior that govern the way they respond to both internal and external issues. Additionally, managers often have a near-instinctual urge to defend their organization from "attack" by outside forces, including community residents and other members of the public. Questions about, say, chemical use at a facility, or expressions of concern about the quality of water drawn from a private well a mile from a contaminated site are sometimes taken personally, or as indications that the person who asked the question or made the comment is an adversary.

The response by an organization to questions or concerns from community residents and other stakeholders will typically be in keeping with overall organizational culture. Thus, organizations that are rigidly hierarchical or that must function within very competitive environments frequently have trouble responding in a positive fashion to questions and concerns from the public.

Managers in rigidly hierarchical organizations have trouble dealing with questions and concerns from persons "on the outside," both because these people aren't in the "chain of command" and because questioning people within the chain of command isn't done from bottom to top—and if community residents aren't in the chain of command, they are likely viewed as being at the bottom. Additionally, organizations that function in very competitive or embattled industries frequently have a "circle the wagons" mentality regarding what they perceive as threats. Such organizations are so used to fending off competitors or other forces that may curtail their operations or cut into their market share that their natural response is either to ignore questions or concerns voiced by community residents or to view them as yet another attack aimed at hindering their ability to operate.

A commitment to community relations typically has to come from the top levels of an organization. Middle managers are rarely in a position to devote time and resources to community relations without the support of senior managers. In addition, since community relations involves providing information on a facility, site, or project to stakeholders, middle managers may be reluctant to do so without authorization and encouragement. After all, who would engage in a dialogue with the community if doing so could imperil one's job? Particularly in the case of serious issues or concerns, the "safest" course of action for middle managers (and even some senior managers) in the absence of organizational support is to do nothing, even though this response may, in extreme cases, lead to the inability to continue operating amid severe public opposition. Thus, it is up to those at the top to understand the need to deal with the public's concerns about environmental issues and to set policies and provide resources that will enable managers in their organizations to work with stakeholders in a positive fashion.

Another cultural norm that can get in the way of meaningful communication and the building of relationships with stakeholders is "professional standards" for communication. Large organizations, in particular, whether they are within the public or private sector, are used to a certain formality in correspondence, meetings, and other activities. They may unconsciously apply the same standards to community relations that they apply to business-to-business marketing, advertising campaigns, or correspondence to members of Congress. These standards may allow for a formal meeting with a community's mayor, but they certainly wouldn't encourage an informal chat with a residential neighbor in his or her backyard as an actual business activity, and, as we will discuss, informal chats with the neighbors can be among the most effective of community relations activities.

Thus, inability to separate organizational communication standards from what may be appropriate for a specific community is a major reason why organizations that understand the need to work with the community fail miserably after taking what could best be described as a "marketing approach." The marketing approach is big on formal meetings with formal leaders and "slick," one-way communication that may or may not address the concerns or answer the questions of specific stakeholders. More often than not, this approach doesn't work because it doesn't bring the right people to the table and it doesn't encourage the input necessary to design a positive resolution to whatever the problem might be.

In a very real sense, persons who are working within the community relations process need to view themselves as functioning as advocates for the community. This advocacy role benefits the organization responsible for the facility, site, or project by ensuring that, while other managers in the organization are focusing on the organization's needs, the community's needs are considered in the decision-making process, as well.

The Business Case for Working with the Public

From a business strategy point of view, the need to address public concerns regarding environmental issues should come as no great surprise. Good strategic analysis of "threats and opportunities" facing a company (or a public sector entity) should consider the effect of the public's response to its operations along with other outside influences. The cost of ignoring community concerns about environmental issues can be high, as companies discover when "the invisible hand" of Adam Smith smacks into the brick wall of community opposition. Good relations with the community constitute an "insurance policy" of sorts that can cushion the organization responsible for a facility, site, or project from the worst effects of crisis situations.

In terms of regulatory and social trends, the genie is already out of the bottle, and the need for developing and maintaining positive dialogues with stakeholders will only continue to grow. Both public and private sector organizations frequently fail to account for the costs of public opposition, focusing instead on the cost of activities such as community relations, despite the fact that proactive communication is almost always significantly less expensive than crisis communication or rear-guard actions attempting to salvage a facility or project into which substantial funds have already been sunk.

As we've said, community residents often become as upset (or more upset) over being "kept in the dark" about a facility, site, or project as they are about the environmental issue itself. Attempting to turn around an angry community and regain its trust after opposition has been mounted can be extremely costly, and is by no means a sure bet. Skimping on community relations in the early stages—whether out of a lack of understanding of the need for such activities, or from a desire to save money—can result in the loss of far greater amounts of money and resources, not to mention public goodwill.

Organizations need to take a different view of the impact of public concern or displeasure and the role that community relations can play in minimizing or eliminating these problems. Costs should be calculated for delays in cleanups, for example, or for the failure to gain a permit to expand a facility. The differences between these costs and the cost of a good community relations program should be recognized for what they are—important cost avoidances.

ABOUT THE INFORMATION IN THIS BOOK

This book is not a scholarly treatise on community relations. Instead, it conveys practical information based on the experiences of the authors and others in both the public and private sectors who specialize in working with community residents to address concerns about environmental issues. The community relations process described in this book has evolved over the past 15 years or so to address community residents' concerns and answer their questions. These concepts and techniques will be useful to anyone who needs to work with the public regarding environmental issues.

Success in addressing stakeholders' environmental concerns typically requires an interdisciplinary mix of skills and knowledge. The ability to communicate and to manage group processes and engage in the building of relationships is one half of the puzzle. The other is sufficient understanding of scientific, technical, and regulatory concepts to promote a substantive dialogue. Since this combination of skills and knowledge is reasonably uncommon, the question is often raised as to whether it is better for technical people to learn communication and group process skills, or for communicators to learn about technical and regulatory issues. The community relations field includes both types of people—and this book is intended for both of these audiences, as well as for facility and project managers who find that they need to work with their neighbors and other stakeholders.

Allied Areas of Study

Persons who are interested in working with community residents to resolve concerns and conflicts regarding environmental issues should also consider reviewing the literature on risk communication—a topic that has been the subject of substantial study over the past 15 years. This book considers risk communication as it fits within a community relations effort; the topic itself is its own specialized discipline.

Two other allied disciplines whose literature is worth reviewing are dispute resolution and mediation, which, for the sake of space, we have not explored in this book. Both of these fields provide additional techniques that can be useful in working with stakeholders to address and resolve community concerns.

MAJOR ENVIRONMENTAL LAWS[1]

Clean Air Act

The Clean Air Act, originally promulgated in 1970, is the comprehensive federal law that regulates air emissions from area, stationary, and mobile sources. This law authorizes the U.S. Environmental Protection Agency (EPA) to establish national ambient air quality standards (NAAQS) to protect public health and the environment. The goal of the act was to set and achieve NAAQS in every state by 1975. This setting of maximum pollutant standards was coupled with directing the states to develop state implementation plans (SIPs) applicable to appropriate industry sources in the state.

The act was amended in 1977 primarily to set new goals (dates) for achieving attainment of NAAQS, since many areas of the country had failed to meet the deadlines. The 1990 amendments to the Clean Air Act, in large part, were intended to meet unaddressed or insufficiently addressed problems such as acid rain, ground level ozone, stratospheric ozone depletion, and air toxics.

Clean Water Act (CWA)

The Clean Water Act is a 1977 amendment to the federal Water Pollution Control Act of 1972, which set the basic structure for regulating discharges of pollutants to waters of the United States. This law gave the EPA the authority to set effluent standards (technology-based) on an industry-by-industry basis

---

1. From "Guide to Environmental Issues," U.S. Environmental Protection Agency, Solid Waste and Emergency Response, 520/B-94-001, August 1994.

and continued the requirements to set water quality standards for all contaminants in surface waters. The CWA makes it unlawful for any person to discharge any pollutant from a point source into navigable waters unless a National Pollutant Discharge Elimination System (NPDES) permit is obtained under the act. The 1977 amendments focused on toxic pollutants. In 1987, the CWA was reauthorized and again focused on toxic substances, authorized citizen suit provisions, and funded sewage treatment plants (publicly owned treatment works (POTWs)) under the Construction Grants Program.

The CWA provides for the delegation by the EPA of many permitting, administrative, and enforcement aspects of the law to state governments. In states with the authority to implement CWA programs, the EPA still retains oversight responsibilities.

Comprehensive Environmental Response, Compensation, and Liability Act (CERCLA or Superfund)

CERCLA provides a federal "Superfund" to clean up uncontrolled or abandoned hazardous waste sites, as well as accidents, spills, and other emergency releases of pollutants and contaminants into the environment. Through the act, the EPA was given power to seek out the parties responsible for any release and assure their cooperation in the cleanup. EPA cleans up orphan sites when potentially responsible parties (PRPs) cannot be identified or located, or when they fail to act. Through various enforcement tools, the EPA obtains private-party cleanup through orders, consent decrees, and other small-party settlements. The EPA also recovers costs from financially viable individuals and companies once a response action has been completed.

The EPA is authorized to implement the act in all 50 states and U.S. territories. Superfund site identification, monitoring, and response activities in states are coordinated through the state environmental protection or waste management agencies.

Emergency Planning and Community Right-to-Know Act (EPCRA)

Also known as Title III of SARA (the Superfund Amendments and Reauthorization Act), EPCRA was enacted by Congress as the national legislation on community safety. This law was designed to help local communities protect public health, safety, and the environment from chemical hazards.

To implement EPCRA, Congress required each state to appoint a state emergency response commission (SERC). The SERCs were required to divide their states into emergency planning districts and to name a local emergency planning committee (LEPC) for each district. Broad representation by fire-fighters, health officials, government and media representatives, community groups, industrial facilities, and emergency managers ensures that all necessary elements of the planning process are represented.

Endangered Species Act

The Endangered Species Act provides a program for the conservation of threatened and endangered plants and animals and the habitats in which they are found. The U.S. Fish and Wildlife Service (FWS) and the Department of the Interior maintains the list of endangered and threatened species. Species include birds, insects, fish, reptiles, mammals, crustaceans, flowers, grasses, and trees. Anyone can petition FWS to include a species on the list or to prevent some activity, such as logging, mining, or dam building. The law prohibits any action, administrative or real, that results in a "taking" of a listed species, or adversely affects habitat. Likewise, import, export, and interstate and foreign commerce of listed species are prohibited.

Federal Insecticide, Fungicide, and Rodenticide Act (FIFRA)

The primary focus of FIFRA was to provide federal control of pesticide distribution, sale, and use. The EPA was given authority under FIFRA not only to study the consequences of pesticide usage, but also to require users (farmers, utility companies, and others) to register when purchasing pesticides. Through later amendments to the law, users also must take exams for certification as applicators of pesticides. All pesticides used in the United States must be registered (licensed) by the EPA. Registration assures that pesticides will be properly labeled and that, if used in accordance with specifications, they will not cause unreasonable harm to the environment.

Federal Freedom of Information Act (FOIA)

The Freedom of Information Act provides specifically that "any person" can make requests for government information. Citizens who make requests are not required to identify themselves or explain why they want the information they have requested. The position of Congress in passing FOIA was that the workings of government are "for and by the people" and that the benefits of government information should be made available to everyone.

All branches of the federal government must adhere to the provisions of FOIA, with certain restrictions for work in progress (early drafts), enforcement, confidential information, classified documents, and national security information.

National Environmental Policy Act (NEPA)

The National Environmental Policy Act was one of the first laws ever written that establishes the broad national framework for protecting our environment. NEPA's basic policy is to assure that all branches of government give proper consideration to the environment prior to undertaking any major federal action that significantly affects the environment. NEPA requirements are invoked when airports, buildings, military complexes, highways, parkland purchases, and other such federal activities are proposed. Environmental assessments and environmental impact statements, which are assessments of the likelihood of impacts from alternative courses of action, are required from all federal agencies and are the most visible NEPA requirements.

Resource Conservation and Recovery Act (RCRA)

RCRA gave the EPA the authority to control hazardous waste from "cradle to grave." This includes the generation, transportation, treatment, storage, and disposal of hazardous waste. RCRA also set forth a framework for the management of nonhazardous solid wastes.

The 1986 amendments to RCRA enabled the EPA to address environmental problems that could result from underground storage tanks storing petroleum and other hazardous substances. RCRA focuses only on active and future facilities and does not address abandoned or historical sites (see CERCLA).

HSWA—the 1984 federal Hazardous and Solid Waste Amendments to RCRA—require phasing out land disposal of hazardous waste. Some of the other mandates of this strict law include enforcement authority for the EPA, more stringent hazardous waste management standards, and a comprehensive underground storage tank program.

Superfund Amendments and Reauthorization Act (SARA)

The Superfund Amendments and Reauthorization Act of 1986 reauthorized CERCLA to continue cleanup activities around the country. Several site-specific amendments, definitions, clarifications, and technical requirements were added to the legislation, including additional enforcement authorities.

Title III of SARA also authorized the Emergency Planning and Community Right-to-Know Act.

CHAPTER

2

THE COMMUNITY ASSESSMENT PROCESS

This chapter explores the community assessment process. In addition to discussing the reasons why the community assessment process should be the cornerstone of every community relations program, this chapter provides information on the types of information that should be gathered (and why), identifying stakeholders, and determining the probable geographic extent of interest or concern. This chapter also provides detailed information on how to perform an assessment, including conducting interviews, reviewing media coverage, and researching other important information on a community. At the end of the chapter are lists of questions that should be considered during the assessment and a suggested outline—with annotations—for preparing a community assessment report.

ABOUT COMMUNITY ASSESSMENTS

Community assessment is best described as a process for systematically examining social, political, and economic issues and dynamics in a community to determine how best to accomplish certain goals and objectives. In environmental community relations, one of those goals will be to engage relevant stakeholders in a positive dialogue about issues that have an actual or perceived impact on the environment and their community.

Assessments may be performed after community concerns regarding a facility, site, or project have surfaced or before residents are aware of an impending issue. If an issue is already the subject of public concern or debate, then the assessment process can be used to gather information directly related to the issue. If the specific issue is not yet public knowledge, however, the process can still yield substantial information that can help an organization and its managers anticipate probable public reactions and concerns. This is done by examining public reactions and attitudes toward other environmental and community issues and by looking at other social, political, and economic factors. Despite the uncertainty that accompanies any effort to understand human behavior, a thorough, well-done assessment can predict community reactions, wants, and needs with surprising accuracy. The key is a systematic approach, such as the one that we outline here.

Although an assessment may focus on the specific environmental issues surrounding a particular facility, site, or project, the process should also involve careful scrutiny of the surrounding community and its residents. For purposes of designing a community relations program, focusing only on the environmental issue at hand is akin to asking a blind man to describe an elephant by touching only its trunk. While community assessments that provide only bare statistics on a community or that narrowly focus on whether elected officials have concerns may provide some useful information, attempting to design a community relations program of any complexity from one would be little more than a shot in the dark.

Managers sometimes assume that they already understand the community and that there is no need to perform a systematic assessment. Although some of these managers may be active in the community, rarely have they stepped back and objectively considered the specific decision-making dynamics or styles and channels of communication necessary to reach and involve all potential stakeholder groups. The authors have never seen a situation in which a well-done assessment has *not* provided new and important information—even in cases in which a facility or project manager has been intimately involved with the community. Thus, even in cases in which facility or program personnel have some familiarity and involvement in a community, the assessment process is still necessary to ensure that *all* of the relevant questions are answered.

Budget considerations are also often raised as an impediment to implementing a community relations program. How much is a community relations program going to cost, and how will its success be measured? The community assessment provides a tool that allows for realistic budgeting as well as defensible information for setting measurable goals and objectives. Assessments also frequently save money on the overall community relations program, since they can help managers pinpoint the most effective activities and concentrate on them. Thus, the assessment helps managers make decisions about the resources they are going to need and provides a framework for measuring the progress of their community relations efforts.

What Is Involved in Assessing the Community?

As we said in chapter 1, neither facilities, sites, nor other projects or issues of an environmental nature exist in a vacuum. The community assessment process should provide the *context* in which the environmental issues in question exist. Thus, a good assessment needs to identify and consider the following:

- The issues (environmental and nonenvironmental) directly affecting a facility, site, or project, and the public's perceptions of them.
- Other community or environmental issues or attitudes that may also affect the facility, site, or project.
- Stakeholder groups and their perceptions and behaviors relevant to the facility, site, or project and other major community issues.
- The dynamics of decision making and power sharing in the community.
- The channels of communication that can be used to reach *all* potential stakeholder groups.

The assessment should provide sufficient information so that the resulting community relations program fits the community, considers all stakeholders, allows for effective communication, and can achieve the program's goals. A list of questions that cover the types of information that a good community assessment should seek to uncover can be found at the end of this chapter. The assessment methodology presented in this chapter is both vigorous and tightly integrated. The categories of information discussed in the following sections include considerable overlap in some areas. This is intentional, for how can one understand why certain segments of a community behave the way they do without considering where they fit within the social and political structure, or what other issues or events, past and present, shape their attitudes, beliefs, and perceptions?

The community assessment process takes a two-pronged approach to ensure a well-rounded and nonbiased view. It typically involves both interviewing a number of persons from the community (or persons who have knowledge about the community or the issue in question) and reviewing documents of various sorts, including newspaper stories, demographic data, and other information about the community or the issue. The framework for community assessments presented in this chapter can be expanded or reduced to meet the needs of a specific issue.

Performing a community assessment is itself an important community relations activity. The act of going to neighbors, key opinion leaders, and other potential stakeholders and asking them for information and suggestions creates a positive first step in establishing a dialogue or reviving/renewing a relationship that has gone awry.

When Should an Assessment Be Performed?

Since community relations programs function most smoothly when they are undertaken as proactive measures, it is best to perform an assessment as early as possible after an issue, or the potential for an issue, has been identified. For example, in the case of a contaminated site, an assessment should ideally be performed as soon as possible after the contamination is discovered or before major investigative or cleanup work begins. In the case of siting or expanding a facility that will be handling hazardous chemicals, producing significant emissions or discharges, or otherwise posing a significant impact (or perceived impact) on the community, an assessment should be performed prior to the commencement of the siting or expansion process. In the case of permitting under the Resource Conservation and Recovery Act (RCRA), an assessment should ideally be performed before the permit application is submitted.

IDENTIFYING THE ISSUES

As Figure 2.1 shows, stakeholders view a facility, site, or project within the context of:

- The specific issues surrounding it.
- Other environmental issues.
- Other community issues.

Thus, the assessment should consider all of these issues, as we will discuss in the following paragraphs.

■ **FIGURE 2.1** Stakeholders View Facilities, Sites, or Projects Through the Context of Other Issues.

Specific Issues Directly Affecting the Facility, Site, or Project

The assessment should begin by identifying the issues directly affecting the facility, site, or project that are also likely to affect, or create concerns among, community residents. Although these issues will be primarily environmental in nature, such as the existence of contamination, upcoming permitting activities, or emissions or discharges, attention should also be given to certain *nonenvironmental* issues that can have an impact on the reputation of the facility, site, or project. Labor disputes, significant layoffs, plant closings, changes in truck routes through the community itself, and development of property adjacent to a site or facility are all examples of issues that are not environmental in nature but can lead to questions or concerns about the environmental issues at hand.

It should be noted that environmental issues are a hot button that is sometimes pushed intentionally by people who wish to mobilize community concern to achieve an unrelated goal. The authors have run across a number of cases in which the environmental card has been played in order to pursue some other agenda. Thus, *all* major issues confronting a facility, site, or project should be identified in the assessment.

In one case, workers at a facility that has been the site of significant management-labor conflicts have a history of calling the state department of environmental quality to report suspected releases or other violations whenever they are displeased with management actions. The state department of environmental quality is well aware that the majority of these calls are placed to create problems for the facility's management; however, they still have to investigate. The frequent investigations have given the facility a reputation in the community as a violator, even though its compliance record is actually quite good.

Additionally, the attitudes, opinions, and beliefs of community members about the facility, site, or project in question should be included in the review of specific issues directly affecting or related to the facility, site, or project. For example, in the case of a contaminated site, the attitudes, opinions, and beliefs of community members about contamination and cleanup activities should be included in the assessment report. Misinformation or erroneous assumptions are important to note as well. Other information that should be collected includes the level of knowledge community members possess about the issue at hand. Finally, information should be recorded regarding efforts being taken by the persons responsible for the facility, site, or project to address or manage environmental issues.

Other Environmental Issues in the Community

Other environmental issues in a community will invariably color residents' perceptions. For example, the presence of a hazardous waste site that has contaminated neighboring groundwater wells can sensitize residents to other potential or perceived threats to their water supplies. Thus, a second site with groundwater contamination can easily attract the attention of the media, which may see a reason or an opportunity to expand coverage of this "trend," which, in turn, can create greater awareness and concern among the public.

In another example, although a single minor spill at a facility that uses or stores chemicals may be shrugged off as an "accident," if the spill is one of several recent incidents at area facilities, it is likely that it will receive substantially more notice, and could catalyze a public debate over whether the facilities in the area are posing an unacceptable risk to their neighbors. Indeed, major, well-publicized accidents involving chemicals tend to heighten overall concern about the presence of chemicals in the community. National stories about spills or accidents may be "localized" by community news media with stories on local "threats."

Concern about environmental issues can have an additive effect in terms of community anger. Comments the authors have frequently heard during assessments are that an area is perceived as "the dumping ground" for everyone else (e.g., it has a number of treatment, storage, or disposal facilities (TSDFs) for wastes), or that all industry in an area—or a certain segment of it—is mishandling chemicals or befouling the air or water. In the case of natural resources, development or use of small areas of woodlands, wetlands, or mineral resources may result in little interest; however, widespread development or use of such areas and resources is likely to result in concerns on the part of at least some community residents as to whether continued development or use is wise or desirable.

As in all of these cases, community residents frequently feel that they "have had enough," and while one single incident may have been "forgiven," or one TSDF or cleanup may have gone unnoticed, stakeholders who already feel besieged may well respond with anger that seems out of proportion with the specific incident or issue in question. This type of situation, in which residents perceive that they have been unduly affected by risks or environmental degradation, drives many environmental justice claims.

Local environmental issues can also create an erroneous frame of reference that will need to be countered during the communication process. For example, if community residents are opposed to or concerned about landfill operations—about disposal—then communication about an operation that involves only solvent recovery will need to emphasize that no on-site disposal is involved. Otherwise, when residents hear that the facility handles wastes, they are likely to assume that at least some of these wastes are ending up in the ground. Residents may then oppose the new facility based solely on that erroneous belief.

Other Community Issues

A thorough assessment should identify and analyze other major issues confronting the community. There are two reasons why nonenvironmental community issues should be examined.

First, some of these issues may indeed affect the facility, site, or project as well as the community relations program. For example, serious political

---

The Importance of Identifying Misinformation

In addition to looking at the context surrounding an environmental issue, persons performing assessments should pay close attention to misinformation about the issue. Since community relations efforts often devote as much time to correcting misinformation as they do to communicating correct information, it's important at the outset to know what area residents and other potential stakeholders are thinking and where they got their information.

---

turmoil can create difficulties in forging lasting alliances with elected and appointed officials. It can also set the stage for an issue, such as a cleanup or impending public hearing about a permit, to be pulled into the political fray. Differences of opinion about land use and development, particularly in the vicinity of a facility or site, can focus attention on the environmental issues at the facility or site that might otherwise have gone unnoticed.

The second reason why major community issues should be studied is because doing so provides a unique opportunity to determine *who* the important players are and *how* decisions are made. Studying major community issues can also provide information on the way community members (or the community as a whole) respond to perceived threats. This is vital information for persons who must develop an effective community relations plan.

STAKEHOLDERS AND THEIR AGENDAS

Since community relations involves building relationships, much of the focus of the assessment process is on identifying and understanding stakeholders. The processes of identifying issues and stakeholders typically occur simultaneously. Often, identifying issues and stakeholders is a "which came first—the chicken or the egg" scenario. The issues will determine who stakeholders are and how they will be aligned within the context of the community, but the stakeholders themselves can also influence the relative importance and shape of the issues. Thus, identification of issues and stakeholders is virtually inseparable.

The concept of "stakeholders" is not always clear-cut. The term is typically used to define persons or groups who have an interest in or could be affected by an issue or a situation. It should also include persons or groups who *perceive* themselves as affected.

In one situation, residents in the vicinity of a contaminated site had accepted the cleanup option of on-site incineration. The project was almost blocked by other residents who lived eight miles away and upwind from the site, however. Although, in actuality, these distant residents would not be exposed to emissions from on-site incineration activities, they perceived that they were at risk, and they were ready to go to the governor to stop the cleanup.

During a community assessment, one should attempt to identify *all* potential stakeholder groups—even though all may not ultimately be involved in the community relations dialogue. The decision as to who should be involved is made during the community relations program planning process, typically according to the degree of interest in or impact from the situation or issue. In order to make that decision, it is necessary to identify the groups and gain at least some understanding of their agendas.

For example, in the case of a minor issue, such as a small, distant site with minor contamination, an effective community relations program may involve only the one or two stakeholder groups that would realistically be interested. These groups may include neighbors and representatives from the county government (e.g., health department, emergency services, agriculture department). Although other potential stakeholders may be identified (such as an environmental group headquartered 20 miles away that has an interest in monitoring the cleanup of contaminated sites in its community), if research into the group's activities reveals that its members are unlikely to involve themselves in a minor cleanup 20 miles distant, this group would not be included in the community relations effort.

Still, it is desirable to know that such a group exists, so that if its members *do* make inquiries about the site, those responsible for the community relations effort will know who they are speaking with and what their interests are. In the case of community relations efforts involving a high-profile issue, the circle of stakeholders that should be included in the dialogue will necessarily be much larger.

Stakeholder Categories

"Stakeholder" should not be considered a rigid concept or a permanent affiliation. Persons considered "stakeholders" or members of a specific stakeholder group in regard to one issue may not be stakeholders in regard to another issue. For example, a person who lives in a particular neighborhood may be considered a stakeholder in regard to a nearby remediation project by virtue of the site's proximity to his or her home. This same person may be considered a stakeholder in regard to a remediation project on the other side of town because the site is near his or her child's day care center, or because the person is a member of an environmental organization that concerns itself with contamination in the community. The wants and needs of the stakeholder groups with which this hypothetical person is aligned can vary significantly from issue to issue.

Stakeholder groups are often formal or easily defined groups, such as elected officials, local business organizations, immediate neighbors, local environmental organizations, and the like. Although formal, standing groups *do* frequently function as primary stakeholders, and should be included on the stakeholder list, additional digging and additional analysis may be necessary to uncover other important stakeholders who may be interested in a specific issue. A case in point is ad hoc groups that are formed specifically to address a particular issue, such as the cleanup of a site, harvesting or extraction of natural resources, or the permitting of a TSDF. Such groups are typically among the most involved stakeholders, but their existence may not be mentioned by members of "formal" groups until the ad hoc group becomes extremely vocal, at which point extra effort will be necessary to answer their concerns and deal with pent-up anger over the situation.

The authors have found that the formation of ad hoc groups tends to be common in some communities and occur rarely in others—something that can be uncovered during interviews or by reviewing back issues of the local newspaper. Formation of ad hoc groups may be a logical response in communities that follow either a very inclusive mode, in which many groups are involved in decision making, (e.g., the ad hoc groups are an extension of a culture of involvement in decision making) or in communities that follow a noninclusive mode. In noninclusive communities, segments of the population that lack meaningful access to the decision-making process sometimes create their own forums, which may include the formation of ad hoc groups, to express their concerns or press for results.

Ad hoc groups are often formed by individuals who can be identified before a group is formed (or shortly afterward) by asking during the community assessment for the names of the people who are either opinion leaders or spokespersons in regard to certain issues. These opinion leaders can then be contacted during the assessment and brought into the communication loop early, before the ad hoc group adopts an adversarial posture. If a more low-key approach to the community relations program is desired, or if the opinion leader in question is unlikely to be receptive to early overtures to engage in the dialogue, these people may not be contacted at the outset of the community relations program. However, persons responsible for the community relations program will be ahead of the game if they know of the existence of such opinion leaders, and are not taken off guard if an ad hoc group is formed.

In addition to local government officials, who are virtually always primary stakeholders, stakeholder groups may also be aligned along the following characteristics:

- Geographic area
- Demographic group
- Business or economic interests
- Ideological or interest group affiliation

It should be noted that some stakeholder groups may be defined according to several of these characteristics (e.g., Spanish-speaking residents of a neighborhood near a site or facility). Care must be taken to determine whether, or to what extent, a particular affiliation or characteristic overrides the others when defining the parameters of stakeholder groups for a particular issue. Note that these are *suggested* criteria by which stakeholders may be defined. Not all will be applicable in any given situation. One of the goals of an assessment is, in fact, to determine by what criteria stakeholders should realistically be defined in a specific community.

Geographic Area Stakeholders are often identified by the geographic area in which they live or work. Residential, industrial, and commercial neighbors of a facility, site, or project should automatically be considered primary stakeholders. Neighbors may be affiliated with other stakeholder groups, as well, and these other affiliations may supersede the "neighbor" designation. As we will explore later in this chapter, the size of the relevant geographic area—and, thus, the extent of a geographically-based stakeholder group—will vary from issue to issue.

Demographic Groups Demographic characteristics that can define stakeholder group affiliation can include race, ethnicity, primary language, income, education, and, occasionally, religion. Although stakeholder groups may be defined along purely demographic lines—groups concerned about environmental justice, for example, or groups that speak a language other than English that will require special accommodations to ensure that they can participate in the dialogue—demographic characteristics may also simply be additional components along with other aspects of stakeholder affiliation. Thus, it is necessary to determine if splitting stakeholder groups out according to demographic characteristics makes sense, or if it is better to treat all neighbors, regardless of household income, ethnicity, race, and so on, as one group. The following case recounts a situation in which splitting a demographically definable group out from the rest of the neighborhood created problems.

In an effort to be sensitive to possible language barriers, a company that wished to provide information to neighbors in a predominantly Hispanic neighborhood prepared a Spanish-language fact sheet for distribution along with an English-language fact sheet for its non-Hispanic neighbors. The Spanish fact sheet was greeted with anger and dismay by many of the Hispanic neighbors—virtually all of whom could read, write, and speak perfect English. The neighbors' negative reaction was, in large part, due to their belief that the company's management had no real idea about who they were, and had simply assumed that they could not read English.

This said, demographic characteristics *should* be carefully considered during an assessment to gain a good understanding of community dynamics. For example, if elected officials provide only the names of long-time residents or persons of specific racial or ethnic groups as potentially interested parties, but census data for the community indicate that a significant number of residents are new to the area, are of another race or ethnicity, or speak a language other than English, then persons charged with implementing the community relations program may have to make some special efforts to reach out to potentially underrepresented groups. Often, groups that are not significantly involved with the formal power structure of a community have their own opinion leaders, channels of communication, and power structures that should be identified to ensure their involvement in the dialogue about the issue.

Thus, if community relations specialists want to involve a local homeowners' association in the dialogue about a contaminated site, but, upon questioning, they find that this group doesn't include representation from certain other demographic groups that are present in the vicinity (e.g., non-English speaking persons, low income/high income, new residents/long-time residents, persons of other racial or ethnic groups), then efforts should be made to involve additional groups or opinion leaders who will provide representation for these people.

"New" versus long-time residents. As mentioned in some of the examples cited in this section, a characteristic that falls within the purview of demographics is the length of time residents have lived in the community. The difference between "new" and "long-time" residents can be very significant in some communities. For example, do "new" residents cluster together in new developments on the outskirts of town and keep to themselves, or do they assimilate into the existing community? Are there racial, ethnic, or economic differences between new and long-time residents? Do new residents have a markedly different view of what the community should be? For example, do they want more development or less development? Do they object to or tolerate "dirty" industry, in opposition to the views of long-time residents? Do they take exception to the local landfill or become more angry/less angry about the discovery of contamination at the site of a long-time major employer in the community? The authors have found that new residents and long-time residents can often represent distinct stakeholder groups with very different expectations and agendas.

Business or Economic Interests Other businesses, or the business community in general, often have an interest in the environmental issues surrounding a facility, site, or project. Although the "business community" is sometimes considered a monolith, this is often not the case. Other business owners may be concerned that problems at another facility, or adverse publicity linked to a contaminated site, can create community concern about *their* operations, as well. Frequently, news about an accident at one facility can lead to questions about activities at other facilities in the area. Additionally, suppliers and customers of a manufacturing facility or TSDF, or of a business that harvests or extracts natural resources, frequently have a vested interest in their customers' or suppliers' public image and environmental performance. Thus, they should be considered potential stakeholders.

In one case, a transportation accident that resulted in a chemical release focused attention on the supplier *and* the user of the spilled chemical, both of whom were located in the same community as the transportation accident. Community and media questions about the accident quickly developed into questions about the storage and management of the chemical at both facilities.

Other economic interests can also affect relationships between a site or facility and certain segments of the community. For example, the presence of a manufacturing facility or a contaminated site undergoing cleanup may be perceived as limiting the marketability of nearby properties that other business interests in the community wish to develop. Thus, some of the strongest supporters of a cleanup can be owners or developers of nearby properties and municipal officials who want to increase the community's tax base by encouraging additional development.

Ideological or Interest Group Affiliation Some stakeholder groups are defined according to ideology or interest in environmental protection or social issues. Groups organized around such ideological concepts or interests may include members from a variety of demographic groups or geographical locations. Environmental organizations may be "mainstream" national groups, such as the Sierra Club, whose local chapters may be active in any number of different issues, or community-based and ad hoc groups that address local issues only.

Environmental groups and individuals who are interested in the environment should, of course, be considered during a community assessment. Environmental groups can be as varied as the people who belong to them. Some groups are interested only in conservation issues, and rarely involve themselves in issues such as chemical use at facilities, or the presence of radon or lead paint in housing. Conservation groups are often the most well-known environmental groups, however. They are often the groups that come to mind when persons charged with implementing a community relations program are looking for "area environmentalists" to involve in the dialogue. Depending on the issue and the local chapter's interests, these groups may or may not be true stakeholders. Since their members are attuned to environmental issues, however, these groups may be able to direct persons conducting an assessment to other local or ad hoc groups that are following a specific issue.

Considering Stakeholder Agendas

It's not enough to simply categorize stakeholders. It is important to step back after stakeholders have been identified and take an objective look at what they want. For example, elected officials in a community with a contaminated site will want the site cleaned up, but they may be equally focused on ensuring that they retain control of the situation for political reasons, or that planned development in the community is not adversely affected by the presence of contamination or by acrimonious outpourings on the part of concerned residents. Thus, persons responsible for the community relations program need to be cognizant of such goals so they can anticipate challenges they may face or assistance they may receive during the community relations effort. As Figure 2.2 shows, different stakeholders will view communication about an issue within the context of their own concerns and experience.

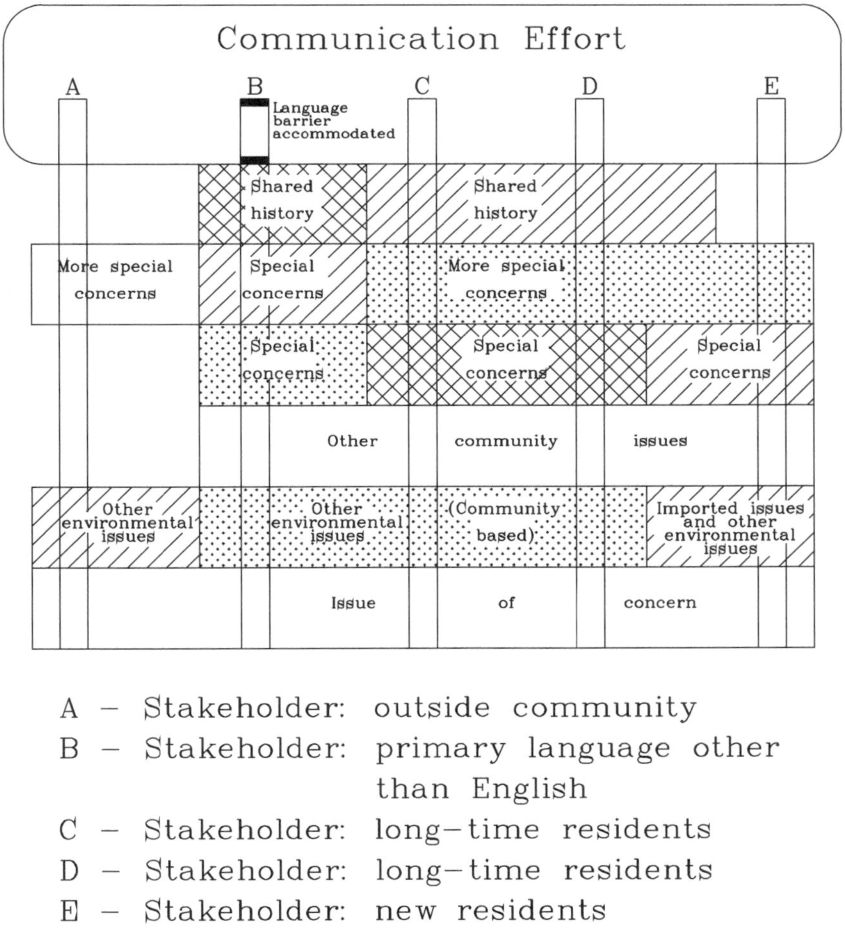

Communication Effort

A B C D E

Language barrier accommodated

Shared history · Shared history

More special concerns · Special concerns · More special concerns

Special concerns · Special concerns · Special concerns

Other community issues

Other environmental issues · Other environmental issues · (Community based) · Imported issues and other environmental issues

Issue of concern

A — Stakeholder: outside community
B — Stakeholder: primary language other than English
C — Stakeholder: long—time residents
D — Stakeholder: long—time residents
E — Stakeholder: new residents

■ **FIGURE 2.2 Different Stakeholder Groups May View the Same Environmental Issue Through the Context of Different Concerns or Historical Perspectives.**

UNDERSTANDING COMMUNITY STRUCTURE AND BEHAVIOR

Although there are certain common patterns of behavior in communities, each community is different. The goal of the community assessment is to identify and make sense out of the many variables in community structure and behavior, and to determine where stakeholders fit, how they interact with each other, and why. The following paragraphs discuss aspects of a community that should be examined during the assessment process. These include:

■ The community's history.
■ Social and political climate and decision-making dynamics.
■ Channels of communication and preferred activities for interaction.

The Community's History

Every community has its own history and, by extension, its own myths about itself. At the most obvious level, historic events of interest to persons performing an assessment will have to do with environmental issues. For example, if the community assessment is being performed in support of efforts to site a landfill, then the history of any previous efforts to site a landfill or facilities for managing wastes should be examined, since community members' recollection of those experiences will influence their initial impressions of current siting activities.

Whether the previous siting effort was successful or not, persons conducting an assessment would be wise to examine both the messages and the actions that the developer undertook, as well as the reactions and observations of community members. Additionally, the person conducting the assessment would also want to look at the experiences that the community has had with existing facilities. For example, if a 50-year-old landfill has contaminated the groundwater, then communication about a proposed landfill would have to provide information on how such contamination can be prevented by new engineering designs and construction methods. Other historical events with environmental aspects should also be considered.

An important part of researching the histories of other waste facilities is analyzing what the community response—or people's recollections of their responses—says about how community members view their roles.

Members of an environmental group in a small town boast about their role in preventing a local manufacturing facility from burning hazardous-waste-derived fuel in an industrial furnace under the boiler and industrial furnace rule, a part of RCRA. They proudly discuss the tactics they used to convince the company not to burn waste, adding that, now that they have forged ties with other environmentalists who helped them defeat the attempt, they intend to continue to stand up to companies that are doing things that, in their view, pose a risk to the community.

The members of the environmental group not only learned that they could successfully thwart an activity that concerned them, but they also learned that they could leverage the resources of their own organization by bringing in outside expertise and assistance—in this case, a group that opposes combustion of wastes. Persons responsible for designing a community relations program for a proposed waste facility in this community should expect that this environmental group will be ready to respond in a like manner, and make sure that they bring this group into the dialogue early on.

Other events of an environmental nature, such as spills or fires at facilities that resulted in harm to the environment, should also be researched to determine what residents remember about them. This provides insight into the types of issues that are likely to be of public concern. For example, a catastrophic fire at a manufacturing facility that has remained in the public consciousness is likely, naturally enough, to sensitize community members to fire hazards. Persons responsible for community relations at a new facility locating in the area should expect that they may be called on to discuss *their* fire suppression system.

Researching past incidents in an area, and then comparing the results to the recollections of community residents, can provide some interesting insights into residents' biases and sensitivities. For example, community members may mention certain issues and omit others, even when both garnered the same amount of attention at the time they occurred. It isn't always possible to track down the reasons for this selective memory; however, when explanations *are* uncovered, they can be quite revealing.

In one town, people recalled—without prompting—the poor environmental record of a metal plating facility, while no one mentioned the poor environmental record of another nearby facility, even though news coverage of fines at both facilities had been about equal and had occurred within the same time frame. Both sites were contaminated and undergoing investigations at the time the assessment was performed.

Further inquiry revealed that the facility with the bad reputation was owned by an out-of-state company and had been the subject of a bitter labor dispute. The other facility was owned by a local family. A number of comments made during the interviews revealed a deep distrust on the part of community members for "outsiders," who were perceived as caring little for the community (e.g., mistreating local workers, harming the environment). Interestingly, contamination at the locally owned site, which was not considered of great concern by the residents, posed a significantly greater risk than the contamination at the former metal plating facility site.

The overall history of an area can also provide some interesting insights into a community and its residents. It is always enlightening to ask community members interviewed during an assessment to describe the main events that have shaped their community—and what effect these events still have today. Asking general questions about the history of the community (as opposed to asking only about specific environmental concerns) often provides incredible insight into the underlying beliefs and attitudes that color residents' perceptions.

Despite increased mobility and rapid growth in metropolitan areas, many communities and neighborhoods have remained remarkably stable. In these places, residents possess a shared knowledge of the area's history (although, in some cases, perceptions of historic events can vary markedly among different segments of the population, as in the case of racial strife). Events such as labor conflicts, disasters of various types, "boom" and "bust" times of rapid development/rapid decline, incidents of political corruption, or the arrival or departure of various demographic groups create a core context in which information about current events is judged.

In the most "positive" of circumstances, community members have behind them a history of cooperation, trust, and some measure of cohesiveness and experience in working together and hanging together during hard times. Historical events in other communities may, instead, have engendered mistrust, deep and bitter divisions between certain population segments, or the perpetuation of an oppressive or corrupt power structure. Any combination of these characteristics will have an impact both on how community residents respond to concerns about environmental issues and how best to work with them during the community relations effort.

The effects of a shared history are typically muted in communities that have undergone significant, recent growth or that have experienced major turnovers in population. History is still important for what it says about "long-time" residents, however. The perceptions of "new" residents and the degree to which they wish to or attempt to learn about the history of the community should also be explored, since they provide insight into how cohesive the two groups—new and long-time residents—are likely to be when confronted with concerns about environmental issues.

An interesting issue can arise when new residents bring environmental concerns with them when they move into an area. In one case, managers of a chemical facility located near a major midwestern city found that new residents to the area—who were primarily migrating from a portion of a nearby city with numerous, serious environmental problems—were much more concerned about the environmental impact of their facility than were the long-time residents. This was understandable, since a number of these new residents had come from neighborhoods adjacent to sites that had been contaminated by industrial activities.

Since the new residents accounted for a significant percentage of the community, the community relations program needed to address the history and attitudes that the new residents brought with them by providing detailed information on environmental controls at the facility and the results of soil testing on the facility property.

Social and Political Climate and Decision-Making Dynamics

An important part of a community assessment is identifying business, social, political, and interest group leaders. These people may be formal leaders of recognized organizations or persons such as elected, appointed, or staff officials, or they may be informal leaders with no formal authority per se, but to whom others look for advice or to serve as spokespersons.

It is important to note that the relative importance of, or power wielded by, business, social, interest group, or political leaders can vary from issue to issue. For example, despite "formal" power, the mayor's views may take a back seat to those of leaders of environmental organizations or homeowners associations if the issue in question is perceived as "closer" to them—and these other leaders are willing to step forward. The community assessment should examine the agendas, behaviors, and relative power of formal and informal community leaders. For example, do elected officials seek input from other community leaders? Which ones? Are some groups excluded from the formal decision-making process? If so, how do they react?

Assessment of the political and social climate of a community involves looking at the other environmental and community issues discussed in the previous sections. Additionally, it requires identifying and assessing the relationships among potential stakeholders in the community. Are all of these stakeholder groups included in the political process, and how do they interact with other groups in the community? For example, are some groups essentially disenfranchised from the political process? Do groups interact socially, or do they form their own "closed societies" with little interaction?

Some communities are extremely cohesive. They face threats (such as contamination) and opportunities (such as grant money) together, either through a directly participative mode or by investing substantial authority in elected and appointed officials or opinion leaders. Other communities are marked by deep divisions, and may respond to threats or opportunities by infighting among dominant groups or by shunting all of the risk onto less powerful groups and ensuring that opportunities benefit the strong.

Modes of decision making can range from highly inclusive, involving many opinion leaders, and community and grassroots groups, as shown in Figure 2.3, to noninclusive, where many groups are excluded from formal

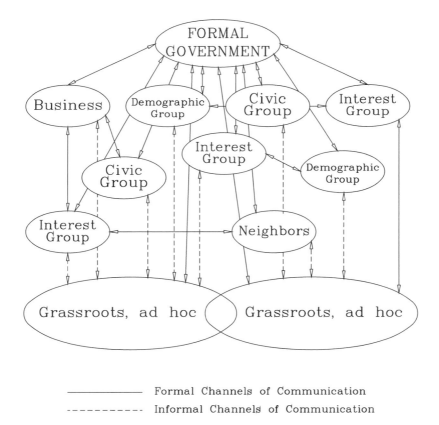

■ **FIGURE 2.3 Stakeholder Groups in Communities with Decision-Making Modes That Are Highly Inclusive Have Both Formal and Informal Input into the Formal Decision-Making Process.**

decision-making processes, as shown in Figure 2.4. In noninclusive communities, excluded groups may either attempt to fight their way into the decision-making process via protests or lawsuits, or sidestep the formal government altogether and pursue their agendas privately on a grassroots level. As discussed in chapter 1, grassroots community groups appear to be on the increase in all types of communities—inclusive and noninclusive. Persons performing assessments would want to look at how these groups are interacting with both the formal government and other business, social, interest, and political groups in the community.

The authors have seen both inclusive and noninclusive modes of behavior in old, well-established, stable communities *and* in new, fast-growing,

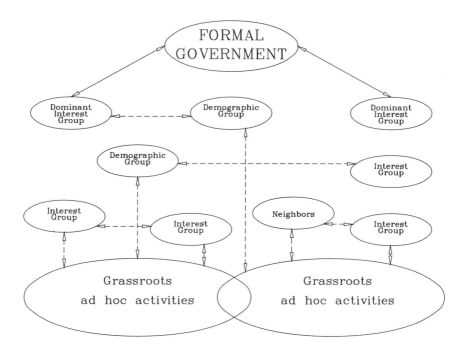

FORMAL GOVERNMENT

Dominant Interest Group

Demographic Group

Dominant Interest Group

Demographic Group

Interest Group

Interest Group

Interest Group

Neighbors

Interest Group

Grassroots ad hoc activities

Grassroots ad hoc activities

———————— Formal Channels of Communication

- - - - - - - - - - Informal Channels of Communication

■ **FIGURE 2.4 Most Stakeholders Groups in Communities with Decision-Making Modes That Are Noninclusive Have Little Input into the Formal Decision-Making Process.**

rapidly changing communities. Although the inclusive mode is often identified primarily with older, well-established communities, it can also be found in fast-growing urban and suburban areas.

Communities that follow an inclusive mode often offer a number of ready-made opportunities for community relations—such as committees, open meetings, and open networks of people who make it their business to know what is going on in the community. Managers who want their community relations efforts to be fairly low-key may find such communities a bit daunting to work with, since it can sometimes seem as if the community relations process is being taken over by the community.

Attempting to implement community relations programs in communities that tend to be noninclusive can be even more daunting, however. For

---

An Important Note About Formal Governmental Bodies

The majority of community relations efforts will include dialogue with formal government representatives as a major part of the program. In addition to looking at how the formal government interacts with other potential stakeholder groups, persons conducting assessments should also consider the amount of power the formal government wields relative to other community interests, such as business or certain other powerful segments of the population. A telling method for determining how much power the formal government has in a community is to look at how the community responds when confronted by a threat, perceived threat, or disaster. Do community members line up behind the formal government—or particular elected officials—or do they avoid formal channels and turn to other business, social, political, or interest group leaders?

Although it is important to identify and work with relevant community leaders, community relations specialists must take care not to undermine the formal government by stepping around them. This can be difficult in communities with strong informal leaders or stakeholder groups and weak elected governments; however, not only is it presumptuous to perform an end run around the duly elected government, but such actions can destabilize the formal government and lead to political strife that can harm the community and hurt the image of the organization responsible for the facility, site, or project that the community relations program was designed to support.

example, the decision makers in such communities often attempt to control the community relations program, dictating who should be involved and what should be discussed. Decision makers in these communities may attempt to keep certain stakeholder groups out of the communication process, saying either that they wouldn't be interested or that they—the decision makers— will act as the sole conduits for information. Intrusion into the community relations program by these types of decision makers can make the free flow of information among potential stakeholders difficult, and can lead to serious problems, particularly if the stakeholder groups that are being prohibited from involvement decide to retaliate by attacking the facility, site, or project that is supposed to be the subject of the community relations dialogue.

Serious divisions in a community can lead to special challenges in the design of a community relations program. For example, does the community have "rich" and "poor" areas? Is there tension among the residents who inhabit them? Are there racial or ethnic tensions? How do these tensions play out? Do these groups attempt to work together and who, if anyone, attempts to mediate?

Channels of Communication and Preferred Activities for Interaction

A portion of the assessment process should be devoted to determining what channels of communication are available and how community residents prefer to express their concerns, receive information, and provide feedback. Both channels of communication and preferences for interaction can be different for different stakeholder groups. For example, the local newspaper may be read only by certain segments of the population. Thus, depending on the groups that must be reached, the local newspaper may not be the best or only choice for dissemination of certain information. Instead, alternate channels of communication must be identified to reach those who don't read the local paper.

In one community, few residents of a low-income neighborhood subscribed to the local paper. Information about an environmental issue—a chemical release at a local facility that briefly affected several neighborhoods—was, instead, successfully disseminated though postings in local grocery and convenience stores, through public service announcements on a local radio station, and through handbills distributed house to house in the neighborhoods in question. A full-page display advertisement providing the same information as in the handbills was placed in the local newspaper to reach the residents in the rest of the community who did subscribe to the paper.

Thus, it is important to ask the people who are interviewed during the assessment for their suggestions regarding the best ways to communicate with community members. Review of newspaper articles can also provide insight into the types of events, such as meetings or open houses, that are customary for that community, locations that are popular for public gatherings, and groups, such as civic organizations, that may co-host, help publicize, or provide a forum for such events.

GEOGRAPHIC EXTENT OF CONCERNS

This book examines building a positive dialogue with "the community," but what is the community? How far afield does one need to go to identify all potential stakeholders? An important part of any assessment is to determine the likely geographic extent of potential concerns about an environmental issue. This step is necessary to ensure that all potential stakeholder groups are identified and that the community relations program is designed to involve those likely to actually be affected by or concerned about the issue in question without having to involve "the world." The geographic extent of potential concern will depend on both the issue itself and the manner in which stakeholders in the area respond to perceived risk.

The Nature of the Issue

The following paragraphs provide a general discussion of how the nature of an issue can affect the potential geographic extent of concern. Table 2.1 provides information on the suggested geographic extents of some common issues. It should be noted that these are *suggested* extents, for discussion

■ **TABLE 2.1** Suggested Geographic Extent of Concerns

Issue	Likely Extent
Physical appearance	Closest neighbors; concern of those who view it
Minor, on-site contamination, no groundwater involvement	Neighbors; persons along truck routes if waste is shipped for disposal; persons located near disposal site

■ **TABLE 2.1** *Continued*

Issue	Likely Extent
Major contamination, off-site groundwater impact	Whole neighborhoods/community; persons on groundwater supply; persons along truck routes if waste is shipped for disposal; persons near disposal site
Contamination of surface water resources and ecosystems	Local and "downstream" communities; possible regional interest; persons who are using surface waters for drinking, recreation, commerce, or subsistence fishing
Major impact to important resources or ecosystems	Local, regional, and potentially, national, interest
Air emissions from incineration, toxic releases	Local and adjoining downwind *and* occasionally, upwind communities. Concern can be very widespread (30-40 miles)
Odors	Extent that odors can be sensed. Since odors are airborne, may catch attention of people outside the range at which they can be sensed
Noise	Extent that noise can be heard
Transportation of wastes or hazardous materials	Neighbors of source/destination, and persons along transportation routes. If "import" of wastes is a hot button, entire communities may be concerned
Spills and accidents	Magnitude of spill or accident will determine whether concern is local, regional, or national

purposes only. Actual extents may vary considerably, depending on some of the other variables we have been discussing in this chapter.

As a rule of thumb, issues such as objections to the physical appearance of a facility or site would be of interest only to those who see the facility or site—those who live or work nearby or, possibly, who drive by. Minor, on-site contamination that does not affect the water supply and that is not near a "sensitive" land use, such as a day care center, is most likely to concern those who work at the site or who live or work very close to it. Such a situation could also conceivably create concerns among local officials or developers who are sensitized to contamination in the community or in areas that they wish to develop.

Contamination that has migrated off-site onto residential or commercial property or that has affected the quality of nearby water wells will have a much greater potential to concern a larger group of people—including environmentalists and neighborhood organizations.

Issues involving resources such as lakes, rivers, and coastal areas have the potential to interest a significant number of people, especially if these bodies of water are used for recreation, drinking water, or commercial or subsistence fishing. People from several communities may want to be involved in the dialogue if they perceive that the issue is affecting them—which is frequently the case with surface waters used for recreation, fishing, or drinking water. If the issue is a major petroleum or chemical spill or a project or development that could have a major impact on the area's ecosystem, then the issue could gain regional or national attention, especially if the body of water is an important wildlife habitat or has other cultural or recreational significance. In such cases, the community relations program would involve dialogue with area stakeholders; however, it would have to be augmented by a program that addressed regional and national media and the concerns of "remote" stakeholders. Such community relations programs would also necessarily have to devote considerable attention to the effect that national media attention and/or the involvement of national environmental groups are having on the dialogue with local stakeholders.

Issues involving other natural areas, such as forests, can also mobilize people from outside the immediate area if these areas are used for recreation, considered important habitats, or have other cultural significance. Again, as in the case of bodies of water, issues involving major changes, harm, or perceived harm to an ecosystem valued for its scenic beauty or wildlife can result in regional or national attention. If the issue involves mining, harvesting of timber, or other such activities, the person performing the assessment will often find stakeholder groups holding a variety of positions on the issue— those who want the activity and view the economic benefits as overriding the desire to keep things as they are, those who are completely opposed to the activity, and those who believe that the activity can be controlled, allowing for both development and conservation or protection of the environment.

Regional or national industry and environmental groups are also likely to weigh in on all sides, adding complexity to the local communication effort.

Issues involving emissions to the air definitely have the capacity to mobilize concerns beyond the immediate vicinity. Other than issues involving widespread degradation or potential degradation to important ecosystems, which, as discussed in the previous paragraphs, can result in national attention, issues involving air emissions typically have the greatest capacity to mobilize concerns over a large geographical area. The potential level of concern over air emissions is understandable—after all, everyone breathes the air.

In one case, public meetings related to an attempt to site a commercial solid waste incinerator in a small town located in a predominantly rural area were attended by people from as far as 50 miles away. The primary concern reported in the press was emissions and the effect they might have on overall air quality and on crops. The turnout of people from miles away was particularly telling, since a community assessment performed in one of the towns that contributed many of the attendees indicated that environmental issues in other communities were considered the "business" of that community, and only rarely—such as in the case of siting a regional landfill, where protection of area groundwater was raised as an issue—did people from outside the community "cross over" to attend public hearings in other towns.

Persons who are designing community relations programs to establish and facilitate dialogues to discuss the "worst-case scenarios" required under Section 112(r) of the Clean Air Act should consider that persons living or working in the area within the extent of the worst-case emissions plume may have an interest in their facility and plan accordingly. In some cases, this may mean talking with people who live a number of miles away.

Odors represent their own challenges. The geographic extent of concern will typically include those who experience the odor; however, odor problems can also heighten concerns about their source among local officials, business persons and developers, environmentalists, and others who believe that the odor is indicative of other environmental problems that could be harming residents or otherwise making an area less desirable to live or work in. Again, since odors are airborne and noticeable, they are more likely to attract adverse attention than minor soil contamination.

Noise from airports, mining, and construction activities also constitute a unique challenge. The persons who are likely to be bothered are those who are experiencing the noise. Since standards have been set for allowable noise levels and noise can be measured and experienced by others who visit the affected area, angry area residents can often "prove" their case and prevail on government officials to take action with much greater ease than in the case of intermittent odors that may or may not be detectable by instrumentation or at a given time of day.

The geographic extent of issues such as transportation of hazardous waste may include persons who live near routes of transportation that wind through a number of communities. Thus, a community relations program would have to consider concerns within a number of different community settings in order to address transportation issues effectively.

Issues involving materials that the general public finds worrisome or frightening, such as radioactive wastes, PCBs, or dioxin, also have the capacity to lead to significant and widespread concern, and, in some cases, national attention. Mention of such well-known, high-profile materials or wastes can easily grab public attention—often regardless of concentration or potential for exposure—and lead to a greater level of concern than the presence of less known substances that, in some situations, may actually pose significantly greater risk to human health and the environment.

Last, environmental hazards or perceived hazards that affect certain segments of the population can lead to greater-than-anticipated interest—if not active participation—in a situation. For example, the environmental justice movement has served to sensitize many minority communities to actual or perceived environmental hazards. News of such hazards in predominantly minority communities can draw support and concern from persons from adjoining communities or regions, or nationally.

In another example, issues involving exposure of "sensitive" populations, such as children, to environmental hazards can also be well-publicized. Such situations can trigger widespread concerns regarding similar scenarios.

A story about a suspected cancer cluster and indoor air quality problem at a public school received national attention—and spawned other "local" stories on the "health" of school buildings in other parts of the country.

Influence of Community Dynamics and Attitudes

Along with the issue itself, another factor that will determine the geographic extent of concerns will be the dynamics and attitudes of the community. Some communities are strictly segmented, with issues involving a particular neighborhood or part of town considered "their business" only. In other communities, such boundaries either don't exist or are frequently crossed.

These attitudes or behaviors can apply between municipalities or other jurisdictions, as well.

When one of the authors performed a community assessment to develop a community relations program for a TSDF, she was told repeatedly by residents of the host community that although *they* were very interested in learning about the facility, the residents of the *adjoining* town—the boundary of which was located less than a quarter of a mile from the facility—would have no interest in being involved in a public dialogue about the facility's operations. Persons interviewed stated very firmly that "we mind our own business here." Interviews in the adjoining town confirmed this attitude—there was no interest on the residents' part in joining the dialogue, despite the fact that residences from their town were located less than a mile from the facility.

Because of the proximity of the facility to the other community, and the fact that some trucks bringing wastes to the facility traveled through the adjoining community on a regular basis, the community relations plan *did* include contact information on the other community in case of an emergency; however, the bulk of the community relations program concentrated on establishing a dialogue within the city limits of the host community.

Another variable that should be considered is community residents' and opinion leaders' willingness to invite in or work with "outside" interests when addressing environmental issues. Some communities can be described as hostile toward "outsiders," such as persons from environmental organizations or persons from other communities or counties who wish to be involved in an environmental issue. A hostile response can also greet outside industry and environmental protection agency personnel. Other communities welcome or call in persons from the outside—although this may be done selectively, with environmental groups welcome (or *some* environmental groups welcome), and agency and industry representatives unwelcome, or vice versa. The extent to which communities allow or encourage involvement of people from "the outside" can be an important factor in the design of the community relations program.

Community biases or attitudes that influence the geographic extent of concern can be readily determined during the assessment process, both by asking community residents and opinion leaders directly who is likely to be interested in an issue (e.g., people on the other side of town, people in an adjoining community, area environmental groups) and by considering the comments they make about the extent of involvement in other environmental or community issues. Review of newspaper articles on environmental and community issues will also provide significant insight into whether community members are likely to recruit or tolerate the involvement of outsiders—and whether attitudes differ toward them if they are from environmental or social justice groups, industry, or the state or federal government.

DETERMINING THE SCOPE OF THE ASSESSMENT

This chapter provides information applicable to conducting an assessment in conjunction with a major issue. While conducting a thorough assessment will no doubt reveal interesting information, the magnitude of the assessment should generally match the size, seriousness, or potential level of concern of the issue.

For example, if the issue involves minor contamination in a rural area that will not have an adverse impact on numerous water supply wells or on special natural resources, such as a wildlife refuge, the assessment may involve talking to only a very limited number of people, such as near neighbors and the county officials responsible for public health or emergency planning. Such an assessment would also include reviewing back issues of the local newspaper to determine whether the decision-making dynamics of the area require that additional people be "brought into the loop" and for evidence of other, similar types of issues that could sensitize a wider range of people.

If there is a question as to whether a particular issue is likely to be of significant concern, then an assessment may be conducted in phases, beginning with a "screening" process involving media and demographic research and a handful of interviews. Using only the "screening" approach, however, may result in a potentially disastrous "false negative" if the community is highly fragmented and representatives of certain groups that may be interested in the issue aren't included among the handful of persons interviewed. To avoid this type of scenario, it is best to use the information gathered during the screening process to determine what other persons should be targeted for another round of interviews, and conduct that second round of interviews before determining that no additional assessment is necessary.

The question is often raised as to the amount of time necessary to conduct a community assessment. The duration is determined by how quickly one can get in touch with people, as well as the time needed to conduct the actual interviews and research, and assemble the results into a form that will allow for informed decision making. A quick screening may take a few days, whereas a full-blown assessment of a complicated, populous area may take several weeks.

The interesting thing about assessments is that much of the important information can be uncovered relatively quickly and easily. The issues and attitudes that drive community reactions are often quite accessible—if one is looking for them and goes about the process in a systematic way. This is not to say that all beliefs, attitudes, or typical modes of behavior are readily discernible at first glance. Often, community members may not themselves be fully aware of, or able to articulate, their own beliefs, attitudes, or modes of behavior, and rarely are they cognizant of how these differ from those found among other stakeholders or in other communities.

Thus, the effectiveness of the assessment process relies on the ability of the person performing the assessment to step back and take an objective look at what is said and why. This analysis phase of the assessment typically takes as much time as the actual research and interviewing. It is often accompanied by some additional phone calls or research, as the person preparing the assessment report confirms his or her understanding of the meaning of the information that has been gathered.

Who Should Perform a Community Assessment?

The question of who should perform an assessment is frequently raised. As a given, persons who perform assessments should be thorough researchers, familiar with the assessment process, and knowledgeable about environmental issues, community organization, and social dynamics. They should also be honestly interested in promoting a dialogue and working with community members so that their approach to the assessment reflects that interest.

Assessments may be performed by outside consultants, corporate public affairs or community relations personnel, or facility, site, or project personnel. Environmental Protection Agency personnel also perform assessments for sites undergoing investigation and cleanup under the Comprehensive Environmental Response, Compensation, and Liability Act (CERCLA).

Several factors should be considered when determining who should conduct an assessment. Large-scale or serious issues benefit from the use of outside consultants or corporate personnel rather than "local" personnel. Large-scale or serious issues would be described by one or more of the following criteria:

- Large and/or highly visible facility, site, or project.
- Facility, site, or project has the potential to pose significant risk (or perceived risk) or other significant impact to the community, human health, or the environment.
- Situation poses a likelihood of significant conflicts between the management of the facility, site, or project and community residents.
- Community in question is complex (e.g., has a large, diverse, and/or rapidly changing population).

Assessments performed to support community relations activities that meet one or more of these criteria need to be thorough, and should be performed by someone not immediately connected to the facility, site, or project at hand. There are several reasons why an outsider[1] should be used.

First, an outsider is more likely to be objective in his or her approach, and, thus, more likely to identify less obvious social trends or potential stakeholder groups. Second, in cases in which bad feelings are already present in the community about the issue in question, an outsider is more likely to get usable feedback from community members and to view news stories about the issue or about the community and its residents with the objectivity necessary for good planning. Interestingly, in the authors' experience, persons interviewed for assessments tend to view outside consultants as neutral parties—even though they are told that the consultants are working for the company or public entity whose facility, site, or project is the object of concern.

Another advantage outside consultants have in conducting assessments is their ability to avoid answering detailed questions from the public during the assessment. This book emphasizes the importance of openness and honesty; however, one of the objectives of the assessment is to determine how best to provide information to community members to ensure a positive dialogue. In some sensitive situations, premature release of partial information can create some of the problems that the implementation of a community relations program is intended to prevent. An outside consultant can answer truthfully that he or she has limited knowledge about the site, facility,

---

1. In some instances, corporate representatives can function as "outsiders" for assessment purposes. If the object of an assessment is a facility that is a major employer in a community, however, residents for the assessment may be less forthcoming about their concerns to a corporate representative than to a consultant or a representative of a government environmental agency.

or project, and that he or she has been asked to perform an assessment to help the managers of the facility, site, or project understand how best to communicate with or involve the public. A corporate or facility representative could not easily avoid answering questions about past, current, or impending issues without appearing evasive.

Assessments for situations of limited scope may be successfully performed according to the framework in this chapter by "local" facility or project personnel, typically with some guidance from outside consultants, corporate, or environmental agency personnel.

TECHNIQUES FOR CONDUCTING AN ASSESSMENT

The preceding sections discussed the major areas that should be considered in order to conduct a thorough community assessment. The activities comprising the assessment process, which include interviews and review of various written records, are typically carried out together, often in several iterations as additional information is gathered. Thus, information gained from preliminary interviews can point the way to issues that can be researched in back issues of the local newspaper—and past newspaper stories or demographic information can lead to additional questions that can be asked during subsequent interviews. In complex communities, several iterations of interviews–research–interviews–research–more interviews can be called for to gain a good understanding of the community and its dynamics.

The following sections provide information on the techniques used to gather information during the assessment process.

Conducting Interviews

It is important to remember, when conducting interviews, that the goal is not simply to determine if people have concerns about or are interested in an issue—"yes" or "no." If the answer is "yes," certainly the person conducting the interview will want to go on, determine why, and identify other factors that may be contributing. If the answer is "no," it is just as important to determine why it is "no." Is it because the person knows about the issue and has had his or her questions answered? Because he or she trusts the persons involved to handle the issue? Because the person does not believe that he or she will be affected by the issue? Or is it because the person being interviewed is unaware of the issue at this time, but comments that he or she makes about other issues indicate that there will be interest or concern once news of the issue is out? Each of these answers may call for a different response in terms of community relations.

Effective interviews can be conducted in person or over the phone. Phone interviews are helpful for obtaining preliminary information. In the authors' experience, phone interviews are also quite effective for talking with professional staff, such as regional planning commission personnel, formal environmental organization personnel, activist organization personnel, Environmental Protection Agency Personnel, or others who are well versed in environmental and/or community issues. These people typically understand what is being asked of them, since they are also often in positions of addressing public reaction to or concerns about environmental issues, and can provide succinct information over the phone.

Face-to-face interviews are especially appropriate when talking to neighbors or local community or environmental group representatives, as well as with local officials and business and civic organization representatives. Face-to-face interviews should build trust and convey sincerity, both of which are important to obtain in-depth information and to get a true reading on the concerns or potential concerns of community members. Another reason why face-to-face interviews can be extremely useful is because they typically provide a greater length of time to talk.

A model for interviews that the authors use involves discussion of the topic at hand (environmental issues or community issues) followed by some small talk, followed by another round of discussion of the topic in question. This model allows a much more comprehensive exploration of views than would an interview that begins and ends with discussion of the topic—as often occurs during a phone conversation. Some of the most insightful comments can occur at the very end of the interview in the small talk that occurs as the interviewer is thanking the interviewee for his or her time. Most people who do not regularly deal with environmental issues and/or community relations cannot simply reel off a list of the things that concern them, the reasons why they are concerned, or what they want done to remedy the situation. They have to discuss the topic, think about it, and come back to the discussion.

Some people may rail against a facility or contamination at a site at the outset of an interview; however, the goal should be to determine not only that they are concerned or angry (or are likely to be concerned or angry, in the case of information that has not yet been released), but what they know (or don't know) about the situation and all of the reasons why they are (or could become) concerned or angry. Identifying the reasons and other factors contributing to anger or concern is vital to conducting the subsequent community relations dialogue.

For example, a person may say he is concerned or angry about operations at a facility. Factors contributing to his anger or concern could be past incidents there or at other nearby plants, the belief that the facility's presence lowers property values, or because a truck from the facility ran over the family dog eight years ago—an incident that has nothing to do with the environment and everything to do with the person's current dislike of the facility.

Learning what people know or don't know is important to determine both the types of information and the level of detail that should be included in communication about the situation.

Conversely, some people may say at the beginning of an interview that they have no concerns about environmental issues or about a facility, contaminated site, or proposed land use in their community, either because these issues aren't top-of-mind, the person doesn't want to "go on record" as saying something derogatory, or because their concerns are difficult to express. Additional conversation often reveals information that provides an understanding of concerns—or lack of them—either as the issue itself is discussed, or through discussion of other similar or related issues. For example, if the assessment is being conducted to design a community relations program to facilitate a dialogue about the permitting of a facility for the management of hazardous waste and the person being interviewed knows little about it, he or she may still be able to discuss his or her feelings about attempts to permit another such facility, or his or her perceptions, concerns, and questions about hazardous waste.

In all of these cases, small talk in the middle of the interview—or interspersed throughout, as the interview moves from topic to topic—serves to build some trust and familiarity as the person being interviewed comes to perceive the interviewer as a person. The break in "heavy" conversation also allows the interviewee to look at the issue from other angles as the subject is reintroduced. This can lead to additional insights that provide the "whys" that are so important.

Residential neighbors are sometimes concerned about talking to a stranger about situations in their community out of fear of reprisal for saying negative things about officials or businesses in the community. The goal of the assessment is to gather information on concerns—warranted or not. Thus, the person conducting the assessment can offer to record anonymously any views that the person has. Such assurances often persuade reluctant residents to talk with the interviewer. The interviewer should take care to ensure that the guarantee of anonymity is carried out, since breaking such a promise flies in the face of what community relations is all about.

Face-to-face interviews typically last from 30 minutes to as long as two hours, with one hour as the norm. Again, the goal is to allow for the small talk that can lead to conversations that reveal attitudes and beliefs that the persons being interviewed may not tie to the issue at hand—or even know at a conscious level that they have. Some give-and-take is often necessary to get at this information. For example, if the interviewer is seeking to determine how different segments of the community interact, he or she can often help the interviewee articulate on the subject by giving examples of how interactions occur in other communities that he or she has worked in. When providing examples about environmental concerns, the person conducting the interview should be extremely careful not to infer a problem, however, lest the person being interviewed assume that the example is indicative of the issue at hand. For example, one would *not* want to ask hypothetically during an interview how the person would feel about having his or her water supply threatened with contamination.

Face-to-face interviews should be conducted at locations that provide the greatest comfort and convenience for the persons being interviewed. Ask the interviewee for his or her preference regarding a meeting place. In the case of local officials and representatives of business or civic organizations, this is likely to be their offices. In the case of business persons, it may be their offices or a neutral location, such as a restaurant. In the case of residential neighbors and informal opinion leaders, it may be their homes, or a neutral location, such as a restaurant or library. Sometimes, several neighbors may want to meet as a group. This can work out well, but be sure to allow enough time for everyone to talk.

The subject of appropriate dress for face-to-face interviews often comes up. The authors believe that it is better to err on the side of formality with local officials, business persons, and civic or social leaders. Business attire conveys respect for the persons being interviewed. Decisions regarding dress are more complicated when it comes to meeting with informal opinion leaders and residential neighbors. Although a business suit might convey respect and be appropriate for meeting with some residents, notably professional or white-collar persons, such attire can also serve to intimidate or put off persons from whom one is hoping to obtain some fairly personal information (e.g., that "suit" doesn't care about us. He/she is just a slick company/government guy/gal). Depending on the community, business casual or even just plain casual may be a better bet for talking to residents.

Last, if persons conducting interviews are asked by someone they interview to provide information and they agree to provide it, they must follow through. Follow-up thank-you notes or calls should be made in most, if not all, cases. This type of positive follow-up is an integral part of the relationship-building process that is key to promoting a constructive and meaningful dialogue.

Persons Who Should Be Interviewed Identifying the right people to interview is part of the challenge of conducting an assessment. One of the first things to determine is which body or bodies of government have jurisdiction over the area in question—municipal, township, and/or county or parish. Frequently, community assessments will involve talking with both municipal and county personnel. In the case of federal facilities, persons conducting assessments will look at communities located near the facility and should deal with the jurisdictions under which they fall. In cases in which a site or facility is located near a municipal or county border, or the issues associated with it are likely to be of concern beyond the boundaries of the immediate municipality or county, then the assessment should also include interviews and research in whatever other jurisdictions are nearby or likely to be affected. For example, persons in a "downstream" community may be interviewed if the issue in question concerns wastewater discharges.

To ensure that the community is well represented, persons conducting assessments should consider interviewing persons from the categories listed below. These categories are provided as a generic guidance to aid in designing the assessment. Not all will be applicable to every issue or in every community.

- Local public officials, including:
  - Elected officials, such as the mayor/county supervisor and/or representatives from the jurisdictions affected by the issue, as well as other elected representatives known to have an interest in environmental protection or community involvement or who have strong ties to stakeholders who are likely to be affected by the issue.
  - Appointed and/or professional staff, such as city or county

managers, planning and economic development staff, environmental and public health staff, and persons who have community or neighborhood liaison responsibilities.

- Public safety officials, including fire department, local emergency planning committee, and hazmat team representatives.
- State and federal public officials, including:
  - State or federal assembly members, senators, and/or representatives from the area in question.
  - State or federal environmental regulatory agency personnel responsible for the program under which the issue will/might be handled and state and federal agency community relations staff.
- Representatives from civic and business organizations and business interests, including:
  - Service and social organizations, such as the Rotary Club, Kiwanis, and so on.
  - Chamber of commerce and local business and industry association representatives.
  - Developers or property owners with interests in the vicinity of the facility, site, or project that is the subject of the community relations effort.
- Neighborhood or social leaders, including:
  - Representatives of homeowners or neighborhood associations.
  - Local clergy or prominent spokespersons among racial, ethnic, and other demographic groups.
  - Representatives from social services organizations, local schools, and parks districts.
- Neighbors, including:
  - Residential.
  - Commercial and industrial.
  - "Special" (e.g., wildlife refuge, school, park), as applicable.
- Interest group representatives and individuals known to be opinion leaders in regard to specific issues, including:
  - Environmental organizations or organizations concerned with issues such as environmental justice, economic development, child safety, and the like.
- Key customers and suppliers.
- Employees and other company/facility representatives.
- Local media representatives.

A Note on Talking to Media Representatives During the Assessment Process

Local reporters and editors can provide significant insight into a wide variety of community issues—especially since some of the most interesting information never makes it into the newspaper. Reporters are keen observers who are typically looking for the same things that persons who are performing an assessment are looking for: What are the issues, who are the players, and why are things the way they are? Reporters and editors are frequently interested in the assessment process and are often willing to help. If the community assessment is being performed in support of a facility or company, then reporters and editors can provide valuable information on how facility management interacts with the media, what they do well, and what they could do better. Such interviews also provide an opportunity to confirm how much media representatives know about particular environmental issues, sites, facilities, or projects and what their attitudes are.

The question is often asked as to whether talking to reporters during an assessment is going to "tip them off" to a problem or issue that has not yet been made public, leading to premature disclosure before managers feel they are ready to answer questions or provide information to the public. If the issue is one of significant sensitivity— or one that has already been the subject of extensive or critical press coverage—then such a scenario is within the realm of possibility. In such circumstances, however, news of assessment activities may get back to the media anyway. Thus, contacting reporters is a judgment call that should be based on the volatility of the situation and likely time frames before disclosure of information about the issue is made.

Most community relations programs will include some work with the media to disseminate information about an issue. Building good relations with the media will be an important part of the community relations program, and interviewing a local reporter or editor as part of the community assessment process can be a good way of establishing that relationship. One word of caution: The person performing the assessment should be aware that there is no such thing as "off the record" in conversations with the media. Although the person conducting the assessment will want to be up-front about gathering information for an assessment, he or she would not want to disclose any information that he or she wouldn't want to see or hear in the news about why the assessment is being done if the reason is especially sensitive.

Locating Persons for Interviews Government officials are typically the easiest to identify and contact. If preliminary work on setting up interviews is being done from a remote location, names and phone numbers of key government officials can be obtained from national directories of municipal and county officials that are available in the reference sections of most public libraries. National directories listing local chambers of commerce and economic development organizations are also often available at the library. Additionally, most chambers of commerce publish their own lists and/or directories of local officials and community organizations for free or for a nominal charge. This information can be ordered in advance by telephone.

Chamber of commerce and economic development organization personnel—as well as public officials—can also frequently provide the names and phone numbers of local business, neighborhood, and social group leaders, as well as representatives of environmental organizations. Once the person conducting the assessment is in the community, review of the area's phone book can yield listings of organizations, including environmental groups, social service groups, and business organizations. During the course of conducting interviews, the person performing the assessment should ask for suggestions regarding additional people who should be interviewed. Referrals are important to locate informal stakeholders; however, as discussed previously in this chapter, "formal" leaders, such as public officials in some communities, may not be completely aware of other potential stakeholder groups or may intentionally omit mentioning groups or individuals that they don't like or with whom they have no ties. Thus, a variety of other methods for identifying potential stakeholders should be employed. Review of newspaper stories about community and environmental issues typically reveals the names of persons and organizations active in environmental, social, or neighborhood issues. These people can often be tracked down in the phone book. The reference desk of the community's library also often keeps lists of community organizations with contact information. If neither of these sources yields a phone number, then another route is to contact the reporter who wrote the story and explain that you are performing an assessment and you would like to get in touch with a person or group that was mentioned in the story.

Finding neighborhood opinion leaders or informal opinion leaders who represent certain racial or ethnic groups, or groups of "new" or "long-time" residents, can represent a significant challenge. Good sources for locating these people include, again, local reference librarians (who, incidentally, often make good interview subjects themselves, since the authors have yet to find one who didn't have good information on the community, its issues, and dynamics), local public school administrators, clergy, and social service agency personnel. Information on neighborhood groups is also frequently posted in local grocery stores. Information on grassroots environmental groups frequently appears in local grocery stores, health food stores, New Age bookstores, and other places—such as bike shops and sporting goods stores—that cater to persons interested in outdoor activities.

Reviewing Media Coverage and Assessing the Function of the Media

Interviews provide some of the most important information gathered during an assessment; however, review of back issues of local and regional newspapers also provides important information. Review of past news stories can confirm or rebut factual information, as well as impressions of social and decision-making dynamics that might have been gained through interviews. Reviewing past news stories fleshes out the information that is obtained through interviews, often pointing up community attitudes that are so pervasive that they are never specifically articulated during interviews, since residents take them as a given. Review of stories also provides additional background that can be used to pose questions during interviews to gain a more complete picture of the community and its dynamics. Although most assessments rely on review of print media, tapes or transcripts from television or radio news broadcasts may also be reviewed in some "high-profile" cases.

If a site, facility, project, or issue has already received media coverage, it is important to ascertain what the public has been told. Critical review of newspaper stories can point up biases in what various persons who are interviewed report and provide insights into those issues or elements of a situation that are considered especially important. For example, if someone reports that an issue isn't important, or that only "crackpots" are concerned about an incident, but a media review of the issue in question reveals significant news coverage that indicates a broad base of concern, that person's perceptions and/or motives should be further examined.

Additionally, review of newspaper stories can provide an idea of how the public views certain issues or organizations. For example, if the issue in question is a contaminated site that is still relatively unknown to the public, review of stories about other contaminated sites can provide insight into the assumptions that community members or specific stakeholders are likely to make about the new site.

In one case, a high-profile contaminated site caused significant adverse impact to groundwater. When another contaminated site was discovered, community residents automatically assumed it would contaminate other area water wells, even though there was no evidence of groundwater contamination. This concern was easy to understand—and anticipate—given an understanding of the previous situation, which served as the point of reference for residents regarding contamination and its effects.

In many instances, the tone of coverage regarding a spill at a facility can provide some insight into general attitudes toward the facility. Is the coverage overly condemning, as opposed to simple reporting of the facts? Are public officials quoted as being especially critical of the facility's management—beyond simple dismay over the incident itself? The person performing the assessment will have to determine whether these opinions reflect primarily the attitudes of the media or attitudes of other persons in the community. If it turns out that reporters are especially biased in regard to certain issues, then this is important information to have, since many community relations programs will involve working with the media as part of the communication effort.

Review of news coverage as a whole—of the types of stories that appear in the newspaper—provides an excellent opportunity for gaining insight into the major issues facing the community. Review of a variety of stories also provides information on who the key players are. The names of informal opinion leaders and information about their agendas can be obtained from news stories. Additionally, as we have suggested, news stories provide a window onto the way the community handles issues and deals with threats or perceived threats.

For example, as we discussed in an earlier section of this chapter, do community members look to formal leaders, such as elected officials, to solve problems, or do they look to informal opinion leaders or political rivals? Who is quoted? Who is providing answers? Do community residents form ad hoc groups, or does the elected government itself encourage the formation of committees to address problems? How inclusive is the decision-making process? If a community is, for example, 50 percent minority and 50 percent white, but analysis of news stories reveals that most of the formal decision making in the community appears to be made by whites, then extra efforts will probably be necessary in the community relations program to ensure that minority residents have equal access to and equal involvement in the dialogue.

Database Searches versus Microfilm   As the previous paragraphs suggest, obtaining a complete picture of all community issues reported in the newspapers as well as their relative importance requires review of complete newspapers, usually on microfilm. Although spending several hours looking at microfilm may lead to a major headache, this approach provides the background on the community as a whole that the community assessment process seeks to achieve. A thorough approach is to look at two or three months' worth of microfilm for each of the past two or three years, as well as "hard copies" of the newspaper covering the two months before the assessment.

Database searches are also extremely useful—but only if the person conducting the search knows the key words to search on. If one is searching for information about a specific facility, company, site, or geographic location, then a database search on these key words is extremely useful. Additionally, searches using generic words, such as "environment," "environmental groups or organizations," "landfill or dump," and the like can yield other good contextual information. It should be noted, however, that if database searches are conducted at the outset of an assessment—before, for example, the person conducting the assessment knows the names of grassroots environmental organizations—significant stories or even entire issues can be missed. For this reason, the authors recommend that database searches be conducted after a review of "hard" copies and microfilm of complete newspapers provide key words and issues for electronic search.

Another shortcut to researching news coverage of environmental and other community information is sometimes available in the form of library subject files. Some libraries regularly clip stories on local issues and other topics and file them according to subject. Review of these files in conjunction with review of microfilm and "hard" copies can provide a very complete picture of news coverage and local issues.

Media Bias   Persons conducting assessments should be on the lookout for evidence of major media biases regarding certain issues.

In one case, when one of the authors was conducting an assessment in a metropolitan area served by two daily newspapers, she noted that one of the area newspapers printed very few stories about environmental issues such as contamination, fines, or dumping, even though interviews and review of the other area newspaper revealed that these were common problems. In one case, the paper in question had only a small story on a back page about a large fine and prison sentence for managers of a prominent local firm convicted of illegally dumping hazardous waste, even though the story received significant attention in the other area newspaper and nationally. According to persons interviewed during the assessment, the newspaper in question avoided reporting stories that its publisher believed could reflect poorly on the business community.

In other cases, the media can be biased *against* industry or government, or give disproportionate coverage to certain types of issues. Thus, although all media reviews will provide important information, persons performing assessments should consider what biases might mean within the context of the community relations program.

Reviewing Other Important Documents and Records

A number of other documents and records can also be consulted to build an accurate view of the community. The following paragraphs describe this information.

Demographic and Planning Information Demographic data and other information on housing, development, industrial and commercial trends, and commuting patterns should be reviewed. Much of this information is included in census reports, which can be found in the reference section of most libraries. Municipal and county planning departments also have this information available, as well as other special studies or reports that may have been done for the community. For example, a fast-growing community may have a "special census" conducted to qualify for additional government funding or to fine-tune planning for development.

Standard census data include information on race, household income, percentage of owner-occupied housing units, length of time in current residence, educational levels, employment category, age distribution, primary language, and other variables that can be used to confirm information provided in interviews and suggest possible parameters for stakeholder groups. Comparison of census tracts in the vicinity of a facility or site with the area in general can indicate whether environmental justice is likely to become an issue.

Information from local planning departments can also provide insight into projected growth or development in an area so that community relations specialists will be able to anticipate the need to expand their programs to include new residents or make other modifications to reflect changes in the makeup of a neighborhood. Zoning and land use maps, which are typically available through the local planning department, are useful additions to the formal community assessment report.

Other Documents Written histories of the community can often be obtained either at the library or through the historical society or chamber of commerce. As mentioned earlier in this chapter, the history of a community can be a powerful factor in community perceptions, attitudes, and opinions. The short histories that are often available at historical societies or chambers of commerce can provide interesting contextual information on the community.

Additionally, if the area has a local Superfund site, then the local public library is frequently the repository for information on the site. This information is available for public review. A community assessment and community relations plan are frequently included among the files. Although some of these assessments and plans focus only on the site in question, and pay little attention to other community issues, others are quite comprehensive and can provide excellent information on the community and community relations activities undertaken there.

PHYSICAL EXAMINATION OF AREA AND PROXIMITY TO SENSITIVE LAND USES

Persons performing an assessment should always visit the facility, site, or other area that is the subject of the community assessment and take a good look at the surrounding area. Although this may sound self-evident, in the case of facilities, in particular, persons who are performing assessments are often so wrapped up in interviews and other research that they fail to scrutinize the immediate area and instead rely on information from the facility manager. Several times the authors have been told by facility managers that there were no homes or schools near the facility—only to find apartment buildings across the street and an elementary school a block away. Facility personnel often don't "see" the areas in which they are located. The same holds true for contaminated sites and the persons charged with investigating or cleaning them up.

Every assessment should include a detailed description of the area around the site, facility, or area of interest, including names and types of businesses and number and types of residences (this can be an estimate). This description should also list other "sensitive" land uses within at least a one-mile, or, preferably, a three-mile radius. This radius can be expanded to suit the issue as described in the section of this chapter on geographic extent. "Sensitive" land uses include parks, schools, hospitals, nursing homes, day care centers, clinics, community centers, nature preserves, churches, shopping areas, and other important landmarks or places where people congregate. In addition to locating such land uses by driving or walking around the area, municipal planning departments often have maps that show the locations of many of these features.

ASSESSING AND ASSEMBLING THE DATA

Regardless of the scope of an assessment, its findings should be recorded in report form. A model outline for a community assessment report appears at the end of this chapter. There are two reasons for this (apart from the fact that, if the work is done by a consultant, the client will expect a report). As described in this chapter, the assessment involves the gathering of an incredible amount of information from many sources. The act of assembling this information into report form requires organization and analysis of the information, which helps the person who performed the assessment identify patterns of behavior that aren't always apparent during the actual information-gathering process. Often, the analysis and report writing phase of the assessment will trigger additional questions that can be asked in follow-up calls to persons who were interviewed. The analysis and report-writing phase often takes twice the time as the information-gathering phase.

The second reason why a written report is important is to provide a comprehensive record of findings about the community. A written community assessment report provides a benchmark of the situations surrounding an issue at the outset of community relations activities and a reference for persons who, several years down the road, may need to know the history of an issue. Facilities typically operate for a long time, so their community relations programs will need an occasional update. Many environmental projects, such as the cleanup of contaminated sites, or the extraction or harvesting of natural resources, can also take place over a long period of time. It is often necessary to revise a community relations program at some point along the way. Reviewing the initial assessment can provide insight into why a program was set up as it was, as well as how a community and its attitudes and perceptions have changed over the years.

An assessment report should provide enough detail so that persons reading it can gain a good understanding of the community and why the person who prepared the report draws the conclusions that he or she has drawn. Since the assessment report is used to help design a communication effort, it should also include as many specifics and as much "flavor" of local ways of talking about issues as possible. The most effective way to do this is to include overall information with a few select quotes in sections of the report that discuss issues, and then include summaries of interviews for each person interviewed in a separate section. This allows readers to review a synthesis of issues, as well as the opinions and comments of specific individuals.

Community assessment reports should also include support materials, such as maps of the area, copies of relevant newspaper articles, lists of public officials, organizations, and individuals interviewed and/or likely to be stakeholders, census data, and any other backup information used to prepare the assessment or that adds to an understanding of the community.

Questions to Consider During an Assessment

- What are the goals of the community relations program? (e.g., to ensure community support or nonopposition for a permit; to obtain input from stakeholders regarding the methods used to clean up a site; to minimize community concerns regarding emissions)
  - Are there any additional organizational goals or objectives that must also be considered?
- What are the specific issues directly affecting the facility, site, or project?
  - Environmental issues? (e.g., history of spills, violations, threat to water supply wells, impact on sensitive ecosystems)
  - Nonenvironmental issues? (e.g., labor strife, rezoning, perceived affect on property values, changes in ownership)
- What are the other environmental concerns or issues in the community?
  - How might these issues affect perceptions or actions toward the issue in question?

- What are other major issues or concerns in the community?
  - How might these issues affect perceptions or actions toward the issue in question?
- Who are the groups or individuals who are affected by or interested in the issue or who perceive that they are affected by it?
  - What links stakeholders to the issue and to each other?
  - Can all stakeholders be reached through "formal" groups, or will special efforts be required to reach certain groups or individuals?
  - What do different stakeholder groups want?
  - What are the demographics of the community, and what bearing might these characteristics have on perceptions of, or interest in, the issue in question?
  - How knowledgeable are the different stakeholders about the issue and about the technical and regulatory issues surrounding it?
- Who are the opinion leaders in the community?
  - What role might these people play in regard to the issue in question?
  - How do informal leaders interact with formal leaders, including elected and staff officials?
- What is the history of the community and stakeholder groups within the community?
  - What does this history suggest in regard to working with stakeholders?
- What are the social and political climate and decision-making dynamics in the community?
  - Is the community's decision-making mode inclusive or noninclusive?
  - How do groups within the community interact?
  - Are some groups disenfranchised from the formal power structure?
  - Is the community one in which ad hoc groups and grassroots initiatives are common?
  - Are "outsiders" welcome or invited into community processes?
- How do community residents respond to threats or perceived threats?
  - Who are the spokespersons?
  - Do residents turn to elected and staff officials or to other community leaders?
- How powerful is the elected government relative to other groups in the community?

■ What are the most effective channels of communication to reach specific stakeholder groups in the community?
  ■ Will several channels of communication be necessary to reach all stakeholders?
  ■ Are there language, cultural, or other barriers that need to be considered to promote a full dialogue?
  ■ Are there any particular types of activities for public involvement preferred by stakeholders?
■ What is the likely geographic area of concern regarding the facility, site, or project?
  ■ What jurisdictions are covered by this geographic area?
■ What sort of media coverage has the issue received?
  ■ What sort of media coverage have other types of environmental stories received?
  ■ Are there any media biases that will have to be taken into account during execution of the community relations program?
■ Is the facility, site, or project located near any residences or "sensitive" land uses?

Annotated Outline of Model Community Assessment Report

I. Objectives

This section lists the preliminary goals and objectives of the community relations program that the assessment is designed to support. Examples of such goals could be to:

■ Reassure community residents that conditions leading to a recent accident have been corrected;
■ Ensure community support or nonopposition for a RCRA permit; or
■ Gain stakeholder input into and approval of a cleanup option for the site.

II. Executive Summary

The executive summary should provide a one- to two-page summary of the findings of the assessment and key recommendations for developing a community relations program.

III. About the Facility, Site, or Project

This section provides a description of the facility, site, or project, its history within the community, current and anticipated issues that are likely to affect its operations or relationship with the community, and a discussion of community relations activities, including formal and informal dialogue with stakeholders, that are currently under way or that have occurred in the past.

Issues likely to affect operations could include labor issues; impending notices of violations; impending expansion; development of adjacent land; discovery of contamination; recent accidents, and so forth.

    A. Physical description and history of facility, site, project, or development

    B. Current and pending issues—environmental and other

    C. Description of current relations with the community

IV. About the Community

This section includes information on all facets of the community that need to be considered in order to design a community relations program. It requires interpretation of the results of both interviews and document review, including demographic data, to identify stakeholder groups, social and political climate and decision-making dynamics, and channels of communication. It also includes an overview of environmental and nonenvironmental community issues.

    A. Description of the community/area surrounding the facility
       Physical description/land use
       Demographics
       History

    B. Issues directly related to the facility

    C. Other environmental issues in community

    D. Other community issues

    E. Social and political climate and decision-making dynamics

    F. Potential stakeholder groups

    G. Channels of communication

V. Interviews

This section includes summary transcripts—paraphrases of general comments, direct quotes regarding issues of specific importance—from all interviews. These summary transcripts should be as complete as possible to allow persons reading the assessment report to understand why certain conclusions have been drawn.

    A. Elected, appointed, and staff officials

    B. Key organization representatives and opinion leaders

C. Neighbors

D. Customers, suppliers, and employees

E. Media representatives

VI. Media Review

This section includes information on news coverage regarding the facility, site, or project, as well as other environmental and community issues. In addition to discussing specific issues and types of issues that have been covered, it should also include a discussion of apparent media biases, level of understanding about environmental issues on the part of specific reporters, and whether coverage of stories in the media contradicts comments about community issues made by persons who are interviewed.

A. Coverage of facility, site, or project

B. Coverage of other environmental issues

C. Coverage of other community issues

VII. Key Issues Identified

This section summarizes the key issues that have been identified in the assessment and discusses their relevance to the facility, site, or project.

VIII. Recommendations

This section should provide recommendations for how the issues identified in the previous section should be addressed. The recommendations should identify stakeholder groups that need to be included in a community relations program, types of information that should be disseminated or developed to answer community concerns and questions or potential concerns and questions, and suggested activities and channels of communication appropriate to reach the stakeholder groups that have been identified.

The recommendations section forms the basis on which a community relations program can be designed.

IX. Attachments

This section should include supplementary materials about the facility, site, or project and the community. Attachments would include maps of the vicinity, census data, copies of relevant newspaper stories, flyers from neighborhood or environmental groups, and so on.

CHAPTER

3

COMMUNITY RELATIONS AND THE COMMUNICATION PROCESS

Communicating about environmental issues is a complex process. Messages, channels of communication, and techniques that work well in one situation and with one audience may be inappropriate or ineffective when transplanted to a different setting. Although one can look narrowly at communication as a way of educating stakeholders about the environmental issues surrounding a facility, site, or project, a community relations program can be effective only if there is two-way communication. Managers must realize that they cannot simply tell the public what they, the managers, think is important. Instead, managers must also listen to stakeholders' questions and concerns and what they expect in the way of action.

This chapter examines the communication process and how its principles can be applied to increase the effectiveness of the communication effort. It also examines three subsets of communication that community relations specialists and managers of facilities, sites, and projects must consider in the course of working with stakeholders: technical communication, risk communication, and crisis communication.

COMMUNICATION IN COMMUNITY RELATIONS

Two-way communication and positive relationships with stakeholders provide the cornerstones of community relations. These two pillars interact to promote trust in the community relations process itself and in the managers and organizations responsible (see Figure 3.1). Communication without good relationships lacks a critical ingredient: As we will discuss in this chapter, stakeholders' perceptions regarding the sender of information can have a strong influence on how the information is received. And relationships require good, two-way communication to weather concerns or potential concerns about environmental issues. Thus, both communication and relationships support and depend on each other to promote a positive dialogue.

As Figure 3.2 shows, failure to establish positive relationships or to engage in two-way communication leaves only one tenuous support—"blind trust"—to address community questions or concerns. Organizations that rely on "blind trust" in the absence of strong relationships and good, two-way communication will find significant community opposition difficult to withstand.

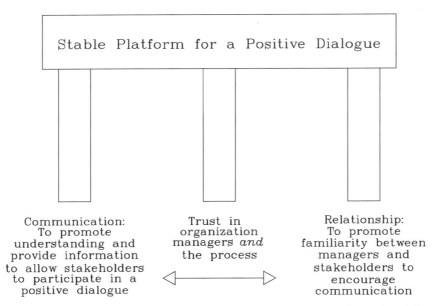

■ **FIGURE 3.1** **Good Communication and Substantive Relationships Create a Stable Platform for a Positive Dialogue.**

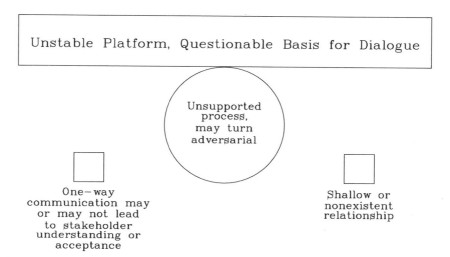

■ **FIGURE 3.2 Little Communication and Shallow or Nonexistent Relationships Provide an Unstable Platform and Questionable Basis for a Dialogue.**

Communication with stakeholders is most effective when it occurs on an ongoing basis as part of a comprehensive, long-term community relations program. Among community relations specialists, the expression is: "Communicate early and often." As we will discuss, proactive communication—early communication—is frequently perceived in a more positive light than reactive communication—even when the same message is given.

As we discussed in chapter 1, the need to establish and maintain positive relationships and positive dialogues with stakeholders is a relatively new concept for many managers of public and private sector organizations. As a result, it often takes a problem, negative publicity, overt public opposition, or a crisis, such as an accident with off-site consequences, to spur managers to consider their facility's, site's, or project's relationship with the community.

Although the initial reason managers engage in a dialogue with stakeholders may be as a reaction to a problem, a proactive approach to communication is ultimately essential to establishing and maintaining a trusted, credible presence within the community. When such relationships are established, they can take much of the "crisis" out of subsequent communication and increase the likelihood that concerns or questions are adequately handled before they escalate.

Goals and Objectives of the Communication Effort

The overriding goal of communication within a community relations program is to provide stakeholders with information that will help them participate in a positive and substantive dialogue with the managers of a facility, site, or project. Stakeholders can use this information either to make suggestions to help inform the decision-making process or to determine for themselves that a course of action is appropriate (or at least, that a facility, site, or project does not pose an unreasonable risk to them).

The goals and objectives of the community relations program's communication effort will depend on the information about the environmental issue that needs to be conveyed. As illustrated in the accompanying box, the goals of a communication effort are what the communicators want the program, or a portion of the program, to ultimately accomplish. What are the desired outcomes? Objectives are intermediate steps along the road to achieving the

---

Goals and Objectives

Problem: Stakeholders are concerned about contamination at a site and are questioning whether or not the investigation will characterize conditions well enough so they will know whether or not they are at risk.

 The following goal and objectives were set specifically to address stakeholders' concerns about whether the investigation will yield good data on the site.

 Goal: To assure stakeholders that the investigation is being properly conducted and will yield good data on the site.

 Objectives: To help stakeholders understand:

■ That the number of samples taken and the locations from which they were collected will adequately characterize the site.
■ How the "analytical parameters"—the chemicals or substances for which the samples will be tested—were chosen, and that the samples will be tested for those substances likely to be present at the site.
■ How the integrity of the test data is assured through the quality assurance process.

---

goals. The likelihood that a communication effort will be successful is increased if goals and objectives are determined at the outset of the communication effort. As we will discuss in chapter 5, clear goals and objectives are also necessary in order to evaluate whether a communication effort has achieved success.

Because of the technical nature of many environmental issues, as well as the emotional reactions that can be evoked by concerns about health risks or damage to wildlife and the environment, it is usually necessary to repeat messages or concepts a number of times, often using several different channels of communication. Setting goals and objectives can help ensure consistency of the messages conveyed within the chosen channels of communication.

Formal, written goals and objectives for communication are also important because they allow community relations specialists to consider how each topic and component of the communication effort fits into the overall community relations program. Additionally, preparation and review of communication goals will minimize the likelihood that important information is omitted, or that the messages will conflict, either with others in the community relations program or with the stated goals or objectives of the organization.

THE COMMUNICATION PROCESS—A BRIEF OVERVIEW

Two-way communication involves both sending and receiving messages. The effectiveness of the communication will depend on how well the message is "encoded" and delivered (i.e., how clearly the message is worded or illustrated and how well the medium used to convey the message does its job), as well as how capable the recipient of the message is of "decoding" or processing the information. When dealing with environmental issues, this basic process, which each of us engages in each day, can pose some interesting challenges. The following paragraphs provide a brief summary of some of the variables that we will be exploring as we examine technical communication, risk communication, and crisis communication.

It is important to keep in mind that, although the thrust of this chapter deals with the considerations involved in "encoding" and sending messages to stakeholders, receiving and "decoding" messages from stakeholders is as important—and, during some phases of the community relations process,

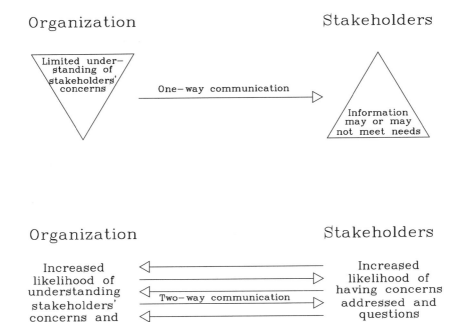

Organization

Stakeholders

Organization

Stakeholders

■ FIGURE 3.3 Two-Way Communication Increases the Likelihood of Successfully Addressing Environmental Concerns.

more important—than the sending of messages. After all, it isn't much of a dialogue if one party isn't listening to what the other party is saying. In such cases, the sender's messages degenerate into a series of non sequiturs that either frustrates receivers into "tuning out" the noise, or angers them into taking action against the sender (see Figure 3.3). Insight gained into stakeholders' concerns, needs, and agendas provided by a thorough community assessment and analyzed during the community relations program planning process can provide much of the context necessary to decode stakeholders' messages accurately.

"Encoding" the Message

There are many issues involved in "encoding" or preparing a message. The following variables need to be considered in the design of a message, whether it is verbal, written, or a graphic depiction:

- The wants, needs, and interests of the stakeholders.
- The level of knowledge or understanding of the subject possessed by the stakeholders.
- The level of detail that stakeholders are looking for.
- Biases or likely emotional reactions to certain types of words (e.g., "toxic," "cancer," "hazardous") or certain types of information (e.g., test results from the analysis of local drinking water).
- Method of presentation and channel(s) of communication (e.g., as a presentation given at a public meeting or as part of an informal chat with the neighbors).

As we will discuss in subsequent sections in this chapter, a significant amount of attention must be paid to the encoding of messages to ensure that they can be readily understood by the people they are intended to reach.

The Medium—and the Spokesperson—Influence the Message

The medium by which a message is conveyed has an effect on the way it is received. Information on environmental issues reported by the news media, for example, typically carries more weight in the general public than information distributed by industry. The type of spokesperson also has a definite effect on how messages will be received. For example, the message that a contaminated site does not pose undue risk to area residents is likely to be viewed very differently coming from a well-regarded university professor or a public health official than it would coming from the owner of the site. Stakeholder attitudes toward different media or spokespersons must be carefully considered during the design of the communication program.

Receiving the Message

There are several important considerations in regard to receiving messages. Persons who are the intended recipients of the message must be capable of receiving or understanding it. This ties back to the process of encoding. For example, use of overly technical terminology creates a common barrier to communicating about environmental issues. Or, if the people who are the intended targets of the message are non–English-speaking, accommodations will have to be made to reach them in a language that they understand.

Stakeholders may intentionally "tune out" information that should be of interest or concern to them if they believe that the communicating organization isn't listening to them. This goes beyond the fact that, if managers aren't listening, the messages are less likely to be targeting community concerns. Intentionally "tuning out" often reflects the lack of a positive relationship and at least a measure of hostility toward those who want to do all the talking, but don't have enough respect for the views of stakeholders to listen to what they have to say. Thus, people have to be *willing* to receive the information that is being conveyed. In community relations, community residents are not *required* to educate themselves about an environmental issue. Instead, it is the managers of the facility, site, or project (and the community relations specialists working with them) who are responsible for reaching out to the community. The best way to promote community interest in receiving information is to make the information part of a dialogue in which residents' input is actively solicited. When stakeholders have had input into the process through interviews (including participating in the community assessment process), small group meetings, surveys, and so forth (or if opinion leaders that they respect have had such input), they are invested in the process, and are much more interested in receiving information because they want to see how the organization is responding to *their* questions, concerns, or suggestions. Additionally, message/messenger combinations must be designed so that stakeholders will be interested in listening to, reading, or looking at the information that is being conveyed.

Assessing the Audience

The level of educational attainment, interests, and preferences of the intended audience for a given communication should be taken into account when determining the following variables that will be used to shape the messages:

■ Choice of language style.
■ Complexity of terminology.
■ Length and complexity of the messages.
■ Types of examples that might be used to explain technical concepts appropriate to demonstrate the information.
■ Preferred format, messengers, and other channels of communication.

The community assessment is the primary vehicle through which potential stakeholders and their concerns and needs are initially identified. In many cases, the assessment will provide sufficient information so that community relations specialists can determine the types of messages, level of detail, channels of communication, types of examples stakeholders can relate to, and so on with a great deal of confidence without performing additional research. In situations involving very serious or complex issues or issues that are likely to persist over a number of years, however, information in the assessment can be used as a starting point to conduct further research to fine-tune the communication effort.

Conducting Additional Research on Audience Needs At this point in the evolution of the community relations discipline, most managers charged with implementing community relations programs do not tend to employ much in the way of formal research. The authors believe that the quality of the communication efforts in many community relations programs could benefit from both the systematic approach and the information that additional research regarding attitudes and knowledge can supply. Thus, we suggest that persons charged with implementing community relations programs should consider using some of the research tools we discuss in this section when designing their communication efforts.

As we said in the previous section, additional research may be warranted in cases of issues that are serious or complex or are likely to persist over a number of years. Issues that involve only a handful of people don't typically require additional research, since much of the dialogue is likely to be conducted face-to-face. Feedback as to the efficacy of the communication effort is usually very clear in such cases. However, additional research is definitely desirable in diverse and/or populous communities, where many people are likely to be concerned about an issue, since much of the communication will necessarily take place through formal channels or activities rather than face-to-face.

Community relations efforts that will require communication with many people can be expensive and time-consuming. Performing some additional research based on the community assessment to home in on specific communication needs or to pretest messages makes sense financially, to ensure that the money and time spent on the communication effort will yield the desired results. Such research can also provide benchmarks to allow the effectiveness of the community relations program to be tracked over time.

Additional research typically comprises gaining more information on stakeholders' levels of awareness and knowledge about an issue and their attitudes toward the issue (e.g., contamination, chemical use, or development of property), the organization responsible, and/or the credibility of potential messengers or channels of communication. Information on levels of awareness and knowledge are important to ensure that messages can be understood. Data on attitudes also provide important information on *what* stakeholders deem important and want to know, as well as information on *how* messages will need to be communicated.

Common tools for conducting additional research about stakeholders' knowledge and attitudes include questionnaires, focus group or small group meetings, and one-on-one interviews.

Questionnaires. Questionnaires can be administered through face-to-face surveying, over the telephone, or through the mail. For community relations purposes, face-to-face or telephone surveying work best, since surveys conducted through the mail tend to have low participation—and those who do participate are often "different" from the people who don't bother to return the questionnaires (e.g., they typically have especially strong feelings on the issue). Thus, extrapolating knowledge or attitudes from a handful of responses can lead to erroneous conclusions.

Questionnaire design is its own art form—and more complex than we can cover in this short space. Mistakes made in questionnaire construction can lead to erroneous conclusions regarding residents' real attitudes. Experts in attitudinal research have found that the order in which questions are asked, as well as the phrasing of the questions, can result in very different answers to what are essentially the same questions. The manner in which a questionnaire is administered, in terms of the consistency with which questions are presented and prompts or explanations are made, can also have a major impact on the validity of test results. Thus, persons who are interested in designing and administering questionnaires should either consult the literature on marketing research or opinion polling for hints on how to design an effective questionnaire or hire a consultant who has expertise in performing such services.

Questionnaires may be administered to a cross section of an entire community or to specific stakeholder groups or subgroups. In cases in which information on the community as a whole is desired, random sampling, in which everyone in the community is equally likely to be chosen for the survey, yields the best results. The more diverse a community is, in terms of either demographics or other stakeholder characteristics that we discussed in chapter 2, the larger the sample size should be to account for the differences in levels of knowledge or attitudes that different stakeholder groups are likely to possess. Information on the diversity of the community should be readily available if a good community assessment has been performed. In very homogeneous communities, a sample of as few as 50 people may be large enough to yield representative data. In diverse communities, it may be necessary to survey several hundred residents to capture the full range of knowledge or attitudes.

Focus groups and small groups. Focus groups or small group meetings are other important tools for gathering additional information to fine-tune understanding of what the issues are, preferences regarding level of detail, desired complexity, and so on. Small group meetings can be as structured or as loose as the situation seems to dictate. A questionnaire administered at the beginning or end of the session—or both—to evaluate the knowledge and attitudes of the individual group members can provide additional information.

Focus groups or small group meetings are intended to provide qualitative information that communicators can use to guide the selection of the content and presentation of materials or messages for an intended audience. Focus groups or small group meetings conducted for research purposes benefit from having an experienced moderator who can guide the exploration of the discussion topics in a manner that will yield specific, usable information. One of the most useful functions of focus groups or small group meetings is to screen out the kinds of materials and messages that are *not* acceptable to most members of the target audience.

In one case, the EPA sponsored a series of focus groups to pretest draft materials explaining the health risks from radon in drinking water. Specific suggestions for improving the message resulted from the focus group sessions, and most were incorporated into the final material developed. As a result of the focus group suggestions, the EPA changed the title of the publication to "Radon and Well Water," eliminated information not specific to private well users, included more information about water testing and treatment, designed a simpler layout, displayed the EPA logo more prominently, replaced the word "mitigation" with a more understandable phrase, and included sources for more general radon information within the material. (From *Communicating Environmental Risk: A Guide to Practical Evaluations*, U.S. EPA. EPA 230-01-91-001. Prepared by Michael J. Regan and William H. Desvousges, December 1990.)

Focus group meetings are typically set up to explore the knowledge and attitudes of, or to obtain feedback from, specific stakeholder groups. For exploratory purposes, however, groups comprising cross sections of stakeholders can also be convened. Typically, small groups or focus groups limited to between eight and fifteen participants are manageable and provide the most conducive atmosphere in which to gather information. The number of focus groups needed to ensure that all of the issues are identified will depend on how divergent the opinions and perceptions of the different stakeholder groups are—information that can be deduced from the community assessment. If there are considerable differences among the goals or agendas of different stakeholder groups (e.g., business leaders, as compared to residents from the neighborhood most affected by the environmental issue in question, as compared to environmental activists), then separate small group meetings with each of these stakeholder groups will be needed to design an effective communication program.

One-on-one interviews. Certain issues or concerns may not be raised in the atmosphere created by a group meeting. Some people may be intimidated by others in a group setting, or they may feel that their comments or questions will be viewed as silly. As a result, it is a good idea to supplement information gained through group activities with some additional one-on-one interviews. The authors have found that one-on-one sessions, though labor-intensive, provide useful insights into the best methods of communicating about environmental issues.

Designing the Message
The content of the message should be based on stakeholders' wants and needs, the channels of communication that will be used, and, of course, on the goals and objectives of the program. Designing a message requires that communicators:

- Compose the message so that the stakeholders will view it as balanced and credible, as opposed to biased and judgmental.
- Choose a language style that is appropriate to the level of stakeholders' understanding, and that takes into account other sensitivities or preferences they might have.
- Set a tone that is neither artificially upbeat nor negative or unduly fear-inducing.

Ideally, the content and presentation of the message should be such that an audience composed of a number of stakeholders of different ethnic, socioeconomic, and educational backgrounds will all view it as balanced and unbiased. Although it is possible to prepare different messages to speak to the concerns of different stakeholder groups, this is a dangerous game, since some may perceive an ulterior motive behind a communication effort that provides different information—or different descriptions of an issue—to different people. Instead, general materials applicable to all stakeholders can be made available (with translations for those who do not speak or read English), with supplemental materials that include more detailed or specialized information available via request or in an information repository. (See chapter 4 for details about information repositories.)

Messages must not appear either too frivolous or too frightening. The audience's reaction can be to ignore or discount a message that appears too lightweight. Worse, an artificially upbeat message may be interpreted as a lie or misrepresentation should subsequent information indicating a larger risk or concern become available. Conversely, one does not want to appeal to fear in order to communicate about an environmental issue or risk. A balanced approach, using clear, simple language, is always best.

Channels of Communication

The community assessment, questionnaires, small group meetings, and one-on-one interviews can all be used to determine the most effective channels for communicating with a given audience. Common channels of communication for conveying information to stakeholders include:

- Print and broadcast media.
- Written materials, such as fact sheets.
- Events, such as public meetings or open houses.
- Spokespersons or "messengers," either from the organization responsible for the facility, site, or project in question or from third-party sources.

The channel through which stakeholders receive information can influence the way information is perceived. This is true in regard to written information, such as fact sheets, distributed by the organization responsible for a facility, site, or project versus information printed in the newspaper, as well as information presented verbally in a public meeting by an organization representative versus the same information presented verbally by a "neutral" third party.

The credibility with which different channels of communication are viewed (including the credibility of certain messengers or third-party sources) will vary from community to community, and often from stakeholder group to stakeholder group. The credibility of various channels available to the community relations effort should be carefully scrutinized, either during the community assessment process or in follow-up research.

Information about the media and the role they can play in the community relations process appears later in this chapter and in chapter 4. In the interest of space, we shall not repeat this information here. Information about written materials and events and how they are used in a community relations program also appears in chapter 4. We will discuss the considerations in enlisting spokespersons or messengers in the communication effort in the following paragraphs.

Spokespersons. The choice of spokespersons can be extremely important when attempting to communicate sensitive information about risk or other factors, such as property values, that affect important aspects of stakeholders' lives. In order to build or support relationships with the community, managers from the facility, site, or project *should* play a prominent role as spokespersons and not hide behind third parties. However, third-party spokespersons, such as college professors, remediation specialists, public health officials, or physicians, can amplify or confirm that the information being provided by the managers is correct. Such "expert" third-party spokespersons can also be called on to explain complicated information, such as the results of a risk analysis.

Other third-party spokespersons who can also help educate stakeholders or confirm conditions at a facility, site, or project can include local emergency response, health department, or environmental department personnel. If these people are not already intimately acquainted with the specific issues surrounding the subject, they should be invited to learn about it. Establishing good working relationships with such local officials should be a high priority of all community relations programs. Not only are local stakeholders likely to call these people if they have questions or concerns, but, in the event of a crisis, the media is likely to call them as well. Thus, their ability to speak authoritatively about the competence of management and conditions at a facility, site, or project is extremely important.

This concept also holds true for elected officials and other local opinion leaders. One of the goals of building relationships and familiarity with formal and informal leaders is to provide these people with the knowledge they need to make comments to others in the community about what they know about a facility, site, or project.

In community relations, the most effective messengers can often be other residents or stakeholders. Comments from such individuals that support and further the dialogue can be more effective than just about any other communication device. As the following example will show, it's important to note that communication does not necessarily have to proceed through a formal leader or "expert" in order to be effective in furthering the community relations dialogue.

A community assessment conducted in advance of a public meeting about the permitting of a landfill included interviews with a number of persons in the neighborhood nearest the site. One of the interviews was with a quiet, yet knowledgeable, man who indicated that he had heard that an "outside activist" was interested in the landfill issue and could show up at the upcoming public meeting. The neighbor did not have a high opinion of the activist, who he believed made money by convincing people to pay dues to "watchdog" their interests. The activist was present at the public meeting and attempted to stir things up. The neighbor got up to speak, announcing that the community did not need outside agitators interfering in its decision making. He said that the agency should have an opportunity to speak and a chance to prove its case. This neighbor's statement made the difference in the meeting, in the ensuing dialogue with stakeholders, and in the subsequent acceptance of the landfill permit conditions.

Addressing Emotional Issues

The kinds of questions and concerns that arise in conjunction with environmental issues can result in a high degree of emotionalism on the part of stakeholders. Strong emotions, whether they are fear, anger, concern, or sorrow, create a barrier to communication that often cannot be breached until the existence of the emotions has been acknowledged. Ignoring emotions or dismissing them as based on misinformation will not solve the problem—it will just exacerbate an already tense situation.

When organizations fail in working with communities over emotional issues, the cause is frequently the failure to acknowledge stakeholders' emotions. "They never *listened* to us," is the usual comment made by community residents in these situations. Indeed, the managers may have listened, but the concept of acknowledging something as intangible as an emotion is so foreign in most organizational settings that managers frequently don't know what to do.

As we will discuss in the section on crisis communication later in this chapter, you *can* acknowledge someone's emotions—their fear, anger, concern—without necessarily agreeing with all of their conclusions regarding its source (e.g., that the presence of contamination at a nearby site is causing cancer). Instead, you are acknowledging—honoring—the fact that *the people* are upset, and that *the people* deserve assistance, information, the opportunity to speak and be heard, compassion. It *is* unfortunate when people are upset—and it *is* appropriate to show compassion and empathy for their emotional discomfort, regardless of whether you believe that their concerns are founded in fact. Empathy is a vital part of effective communication in community relations. Empathy conveys interest and involvement in community residents' feelings and concerns and a desire to fix the problem or to help them understand what an issue means to them.

Empathy must be sincere in order to work. Empathetic statements must be free of attempts to downplay or dismiss the emotions in question. Fake empathy, such as telling community residents (who may be complete strangers), "I know how you feel," is, in many ways, as dismissive as telling people that they don't need to be concerned, since, "I know how you feel" essentially minimizes, on a very personal level, the emotion that the other person is feeling. Most often, it is better not to pretend complete understanding, but instead to simply acknowledge that residents have the emotion through statements such as, "I can see that this is very upsetting to you," or "I can see that you are very angry."

After concerns have been acknowledged—and that acknowledgement accepted—it is often possible to move on to discuss alternative conclusions to residents' beliefs, correct misinformation, or explain the actions that are under way or under consideration for investigating or remedying the situation that has caused the concern. It is important to note that people who are highly emotional are less capable of processing and understanding information than people who are not having an emotional reaction. Thus, when dealing with highly emotional issues, it is best to acknowledge the emotion and then work on conveying only a few key points (or even just one key point). If more complex information needs to be communicated (as it usually does), it is best to do so later, when residents are better able to process the information.

If possible, it is best to deal with emotional issues in one-on-one or small group settings. In some cases, one-on-one and small group meetings can be used for carefully exploring residents' emotions. For example, what lies behind the anger or the fear? Careful questioning can help the meeting participants explore their concerns, and it can create a basis for a less emotion-laden dialogue to explore factual information or options for addressing the issue in question. In order for this approach to work, however, some trust must typically have been established between the person running the meeting and the meeting's participants.

Outside facilitators or third-party experts can be helpful in dealing with issues, such as human health, that have a high emotional content. Outside experts who are able to explain the conclusions of a risk analysis, for example, may be able to bring a measure of comfort to those who are skeptical regarding the motivations of a government agency or a business that has responsibility for the issue. A third party can often bring a credibility that a business or agency spokesperson cannot match.

Addressing Emotion in Large Group Settings   Public meetings are notorious for bringing emotions to a fever pitch. Group dynamics can act to amplify the energy in the room, and the presence of one or two highly emotional individuals can be enough to set the stage for an unpleasant and unconstructive event. The community assessment and other interviews specifically aimed at exploring a particular issue can provide insight into the topics that will come up and the people who are likely to raise them. Whenever possible, these issues should be addressed and resolved before the meeting; however, it should be noted that some individuals who are very upset may be unwilling to meet and discuss their concerns beforehand.

Some commonsense steps can be taken to alleviate some of the tension in a potentially contentious meeting and to prevent emotions from overwhelming a public meeting. In addition to meeting with people beforehand, the authors and other community relations specialists emphasize that thorough preparation, including the crafting of readily understandable answers to questions that are likely to be asked, is essential prior to public meetings that are likely to be emotional. It is also important for communicators who are involved in such meetings to be well versed in, and understand, the nature of group dynamics. A good example of managing group processes in a constructive manner despite the presence of high emotion is illustrated by the following case.

A public meeting was called in a town to discuss a plan that involved digging up the yards of many residences in order to remove soil that was contaminated with radioactive isotopes. The meeting was set for the agency to explain the criteria for removal to a very emotional audience. This was a meeting that had every potential to go poorly, but instead was highly successful.

The agency made a number of good decisions that resulted in a positive experience. First, all of the agency personnel involved worked together, did their homework, and prepared a presentation that made sense. They had the backing of senior-level personnel, who demonstrated their approval by their presence at the meeting. The agenda was clearly set and the purpose and objectives of the meeting were written on a flip-chart for everyone in the audience to see. Before beginning the meeting, the moderator asked for questions and comments to allow the audience to air their anxieties and release their emotions. The audience asked a number of questions and expressed concerns that were all recorded on the flip-chart. The agency responded to the questions honestly and openly so that a positive dialogue was established in the room. The agenda for the meeting was slightly altered, based on the issues and questions recorded on the flip-chart, and senior-level personnel participated in the response. The audience left the meeting in a positive mood and many thanked the agency personnel for their efforts.

Thus, even in a large group meeting, emotionalism *can* be directed in a positive way when the audience's concerns are openly acknowledged (written down, responded to) and people are given an opportunity to actively participate. The communicator needs to take care to engage in "active listening" by reflecting comments back to people to ensure that both parties understand what is being said. Good body language is also important. Engaging in eye contact, moving out among people in the meeting rather than standing behind a lectern, and leaning toward the speaker to listen more closely all work together to show the communicator's respect for and interest in the audience members.

COMMUNICATING TECHNICAL INFORMATION

Communicating information about environmental issues, whether as part of an ongoing community relations program, or in response to concerns about risk or a crisis situation, usually involves providing information about complex technical concepts. The authors have found that it *is* possible to convey complex concepts to laypersons, and that the greatest impediment to the success of such communication often comes from other technical and communication professionals who believe it can't be done. In the authors' own practice, we have specialized in explaining complex technical concepts to the public, and we believe that doing so can often defuse many concerns—especially those based on lack of information and misunderstandings.

Providing interested stakeholders with good information is at the heart of community relations. As Figure 3.4 shows, stakeholders' concerns over an organization's failure to provide them with information can frequently outweigh the concerns they have—or would have—over the actual issue. In cases in which concerns are running high, providing people with good technical information can give them a level of comfort concerning the operation of a facility, site, or project and allow them to convince themselves that actions are being taken, a facility is appropriately regulated, a variety of contingencies have been considered, and so on. Providing stakeholders with good information can also enable them to make helpful suggestions or participate more meaningfully in the decision-making process in cases in which such involvement is desired or required. As we will discuss in the section on risk communication, giving stakeholders a sense of control—which information can provide—often reduces community opposition simply by including stakeholders "in the loop."

The challenge of communicating technical concepts lies in making sure that the information is simple enough for the layperson to understand but still retains its accuracy. Although most members of the "general public" know

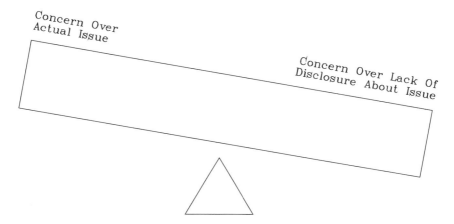

■ **FIGURE 3.4 Frequently, Concern over Lack of Disclosure Outweighs Concern over the Actual Issue.**

little about environmental science, regulation, or technology, there *are* knowledgeable people out there. In addition to people who actually work in the environmental field who may also be community residents (and can sometimes pose the toughest questions at public meetings), many activists are also extremely knowledgeable. In some cases, activists' knowledge can rival that of industry and government personnel. If such people are concerned about a facility, site, or project, they will pick through information released about it with a fine-tooth comb. Information that has been simplified in a manner that could lead to misunderstanding will be attacked—along with the motives and the credibility of facility, site, or project management. Thus, technical and regulatory information prepared for the layperson should be carefully reviewed with the potential for misinterpretation in mind.

It is important to note that, in cases of only mild interest and little concern about an environmental issue, communication can often be simple and brief and still garner community approval. In such cases, basic descriptions of the features or actions taken to protect the environment (e.g., "Leaks from our aboveground tanks would be caught and held in the concrete containment basins surrounding each tank," "We will dig up the contaminated soil and take it away for disposal") provide sufficient information to give residents an adequate level of comfort that a situation is being managed with care.

In other cases, a much more rigorous approach to educating stakeholders may be required. If a significant amount of information will be required to help people understand the technical or regulatory aspects of an issue, then it is best to present the information over time, if possible, so that stakeholders are able to absorb it.

In technical communication, two of the most important factors that need to be considered (along with the many other factors we have already discussed regarding communication) are terminology and level of detail.

Terminology

Technical terms can easily hamper attempts to communicate about environmental issues and, as a rule, should be used sparingly. In cases in which stakeholders have moderate to strong concerns, however, the authors recommend integrating some of the more common technical terms relevant to the environmental issue into the information in a manner that will allow the audience to learn what they mean. We believe that, in situations in which considerable debate over an environmental issue is likely, it is important to familiarize community members with terms that they will be hearing to help them participate more fully in the dialogue. Doing so will allow stakeholders to have more confidence in the decisions that are made or conclusions that are reached—or to participate in a dialogue to inform decision making—because they will have a better handle on what "the experts" are talking about. For example, if the issue in question is contamination of groundwater, then common terms that describe groundwater and geology should be provided and explained. Introducing technical terms has to be done slowly and carefully, and fact sheets or other written materials that use them should contain glossaries with definitions that the layperson can understand.

Level of Detail

Stakeholders can be introduced to key concepts without overwhelming them with more detail than they might want. For example, in one case that the authors worked on, questions were raised regarding the investigation of a site with suspected contamination. Many of the stakeholders were concerned about the way the investigation was being conducted. They had no frame of reference, no experience in such procedures. In order to ensure that residents would accept the data that would be generated by the investigation, it was important to provide them with enough information so they could feel comfortable that the investigation was conducted with care and according to some credible guidelines.

As it was, relatively little detail was necessary to promote confidence in the investigation. The details that were used were as follows:

- The investigation employed standard methods approved by the U.S. EPA for the collection and testing of samples.
- Descriptions of the sampling and testing procedures were included in a formal, written "work plan," which was submitted to the state department of environmental quality for its comment. No samples were taken until the work plan had been approved by the state. The work plan is available for community members to review.
- A state department of environmental quality staff member was present to oversee the sampling.
- The number of samples collected exceeded the number usually collected at similar sites to increase the likelihood that contamination—if present—would be detected.
- The samples were tested for those chemicals that, based on the history of the site, could be present. The state department of environmental quality had reviewed the list of chemicals for which the

samples would be tested and confirmed that the appropriate tests were being conducted.

■ All laboratory tests were conducted under strict "quality assurance" procedures developed by the U.S. EPA to ensure valid test data.

This information is extremely routine and discloses nothing surprising or unusual to people who work in environmental consulting; however, had it been omitted, stakeholders would have been asked to rely on "trust" or, as Figure 3.5 shows, the "black box" approach (the "black box" being the place where "something miraculous happens" and the solution pops out).

In this case, the relationship between the stakeholders—most of whom were residential neighbors—and the organization whose property was being investigated was extremely adversarial. Thus, in the absence of information to provide a measure of confidence in the investigation, the situation would likely have evolved into a major confrontation. In this particular situation, however, the state department of environmental quality was well regarded,

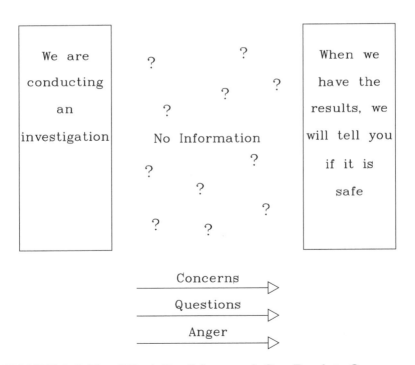

■ **FIGURE 3.5 The "Black Box" Approach Can Escalate Concerns Among Stakeholders.**

and by providing fairly minor detail about its role in the investigation, along with enough other basic information that showed a clear and methodical approach, concerns about the way the investigation was being conducted basically evaporated.

In terms of level of detail, when preparing information, it is best to identify the key concepts or questions that need to be discussed, and then add layers of detail. Additional detail, beyond the first or second layer, can be added if necessary, as more is learned about the interests or concerns of stakeholders. Information containing significantly more detail than most community residents may want can be placed in an information repository for the facility, site, or project or otherwise made available to stakeholders upon request, so that those who want supplemental information can obtain it.

A good way of developing thorough information in a format that is easy to present is to use the question-and-answer format. Questions that people are likely to ask (or that they will eventually ask as a situation proceeds) should be written down and answers prepared. As the exercise progresses, additional questions, additional answers, and additional points that are likely to be important to stakeholders are identified. Questions and answers can be useful tools in preparing for one-on-one meetings or formal public meetings, as well. The authors also use this format to prepare other written materials, since the "Q's and A's" readily break down complex concepts into manageable pieces. The information can be either presented in the "Q-and-A" format or massaged into a logically flowing narrative.

Testing Readability

Readability of material can be evaluated using established formulas designed to determine the grade level a person must have achieved in order to understand the communication. Determining the readability of the material cannot determine its ultimate effectiveness, but when used along with other pretesting methods, it can provide a good indicator of whether it will be understood by the intended audience.

For example, the EPA pretested an early draft of a booklet about lead in drinking water for distribution to a target audience. The pretests revealed that the draft was written at a level more appropriate for managers of the water supply system, did not clearly convey the important message that testing was the only way to determine whether there are high levels of lead at the household tap, and did not give people adequate information about how to get their water tested. (From *Communicating Environmental Risk: A*

Guide to Practical Evaluations, U.S. EPA. EPA 230-01-91-001. Prepared by Michael J. Regan and William H. Desvousges, December 1990.)

Readability formulas use counts of language variables, such as the number of syllables in a given number of words and the length of sentences, to measure the structural difficulty of the written material. As one might expect, the educational level required to understand a written communication increases along with sentence length and number of polysyllabic words. Readability formulas cannot measure the accuracy, organization, bias, or credibility of the communication, however. For this reason, readability should not be the only criterion used to judge the appropriateness of the written material.

Although we have discussed written materials and materials prepared for verbal presentation, graphics can also be extremely effective in explaining difficult scientific and technical concepts. Basic concepts, such as a depiction of how secondary containment would catch a spill, to more complex concepts, such as the operation of certain technologies to clean up groundwater contamination, can all benefit from simple graphics.

PRINCIPLES OF RISK COMMUNICATION

Risk communication is a discipline that has been extensively examined and interpreted by many experts and from many points of view. This section is not intended to add to this cumulative knowledge. Rather, it attempts to interpret some of the general concepts of risk communication as they apply to the communication efforts within a community relations program.

The concepts used in risk communication provide a foundation for those used in many kinds of communication in a community relations program. Risk communication, while a unique discipline, offers a window into some of the issues, problems, and concerns that can arise in all communication involving environmental issues. An understanding of risk communication is indispensable to providing an awareness of how other types of communication can be made more effective.

Defining Risk Communication

Risk communication means different things to different people. The context of risk, for example, whether one is talking about the hazards of smoking or the presence of contaminants in drinking water, plays a significant role in the public's response. The EPA provides a rather compact and manageable definition of risk communication (essentially derived from a

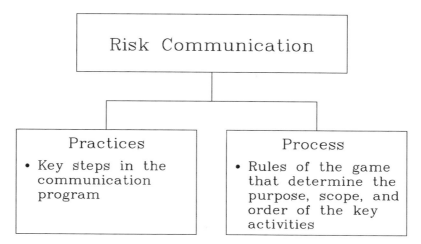

■ **FIGURE 3.6 Dimensions of Risk Communication.** *Source: Communicating Environmental Risks: A Guide to Practical Evaluations.* U.S. Environmental Protection Agency, EPA 230-01-91-001, December 1990, p. 1.

definition by Covello, von Winterfeldt, and Slovic) as ". . . the purposeful exchange of information between interested parties about environmental risks."

Figure 3.6 divides risk communication into practices and process. Practices are the specific steps taken to communicate about risk, such as assessing the audience, determining the message, selecting the appropriate technical language and level of detail, and deciding on the best channels for communication. These steps have already been discussed in this chapter. This section focuses on the process of risk communication and the issues involved in creating a successful dialogue to ensure that a meaningful exchange of information can occur.

The "seven cardinal rules of risk communication" (which also apply to other communications within a community relations program) were developed by Vince Covello and Frederick W. Allen (with the assistance of others in and out of government) and published in a pamphlet by U.S. EPA in 1988 (U.S. Environmental Protection Agency, OPA-87-020, April 1988). These rules are as follows:

1. Accept and involve the public as a legitimate partner.
2. Plan carefully and evaluate your efforts.
3. Listen to the public's specific concerns.

4. Be honest, frank, and open.

5. Coordinate and collaborate with other credible sources.

6. Meet the needs of the media.

7. Speak clearly and with compassion.

Each of these rules plays a part in this discussion and is incorporated through-out the topics presented in both this chapter and the rest of this book.

The Perception of Risk

As we explained in chapter 1, in regard to environmental issues, perception equals reality when it comes to determining whether stakeholders' concerns about risk need to be addressed. If the community perceives that a facility, site, or project poses undue risk, whether or not that perception is based on factual information is not nearly as important as the fact that the perception exists. Failure to acknowledge such concerns—or, worse, brush-ing them off as nonsensical—often serves only to fuel public distrust or anger that can quickly become a major problem in its own right.

Who Should Communicate About Risk?

As we discussed earlier in this chapter, the source of messages about risk—or the person conveying this information—will have a significant effect on how the messages are received and perceived by the community. Risk communication specialists note that the technical content of communication is often not nearly as important to the public as the perceived motivations of the sender.

For example, the person responsible for a risk communication effort might be an industry representative, government official, scientist, or local community member. Each message conveyed will be received by the commu-nity with certain biases and expectations, depending on the communicator's perceived motivations. Additionally, each member of the community will receive the information and process it based on his or her own background and experiences. A message may be totally factual and unbiased, but depend-ing on whether the receiver is a member of, for example, an industry associ-ation or an environmental organization, he or she will filter that message through a set of biases and experiences that will affect how it is ultimately received.

Different residents may perceive the same information in very different ways. This was clearly illustrated by comments made in interviews conducted during a community assessment. One neighbor of a large manufacturing facility (the subject of the assessment) extolled the safety and cleanliness of the facility, saying that he understood from a friend who worked there that employees were provided with uniforms, which were laundered by the company, and that they were required to take showers at the end of their shifts as part of the plant's health and safety program.

Another neighbor had heard the same information; however, he viewed these practices as anything but positive. In his view, making employees wear uniforms and take showers simply proved that the chemicals stored at the facility were "extremely deadly," and, thus, posed an immediate threat to area residents.

As in the case of other communication supporting a community relations program, community relations specialists must ensure that all potential stakeholders are considered in the communication of risk. This includes those persons who are directly affected by the risk being discussed, as well as those who *perceive* that they are affected by the risk.

Barriers to Effective Risk Communication

There are a number of potential barriers to effective communication of risk, and even the most well planned communication may go awry. The community assessment process discussed in chapter 2 is one of the best ways of identifying potential barriers early, so that they can be addressed before the communication effort proceeds.

Both emotions, which we discussed in a previous section of this chapter, and the issue of perception regarding the motivations of organizations or spokespersons *and* the receivers of messages, which we just discussed, present potential barriers to effective communication. Additionally, many people have become oversensitized to environmental issues and the controversies over the effects of chemicals and other exposures that confront them nearly every day. They hear so much conflicting and sensationalized information that it is often difficult for even the most informed members of the public to know what to think about a particular issue.

If people have already made up their minds that something presents an unacceptable risk, then considerable effort may have to be expended to alter these beliefs (indeed, in some cases, it may be impossible to change their views). These efforts may include small group meetings; one-on-one discussions; written information, such as fact sheets; and involvement of well-respected third-party sources. Since community relations concepts are still relatively new to managers of many organizations in both public and private sectors, they may fail to use these alternate means to communicate with stakeholders. Instead, they may attempt to use the most well-known community relations practice—the large public meeting. Large public meetings—in the absence of other, earlier dialogue with the community—are among the least appropriate venues to either dispute or change stakeholders' perceptions or opinions regarding risk.

Communicating Uncertainty   Attempting to explain uncertainty to non-technical stakeholders can be one of the most challenging aspects of communicating about risk. Scientists involved in the assessment of risk are understandably reluctant to offer absolutes, since their studies are nearly all based on a gradation of confidence in a possible outcome. Stakeholders often do not understand why officials may be unable to answer, unequivocally, "yes" or "no" to seemingly simple questions such as, "Is it safe to drink the water? Am I in danger living next to this contaminated site?"

Risk analysis can assess a hypothetical person—say someone of a certain age, who will live a specified lifetime in a certain area—and determine an *average* level of cancer risk. Still, the question arises for the individual: "Is it safe *for me* to drink the water? Am I going to be the one person in 100,000 who gets cancer?" One way of responding is to confront the uncertainty directly and explain why it exists. Rather than increasing the level of anxiety, people can often accept such explanations. What people are unlikely to accept is information that simply confuses them or that they view as misleading.

In cases in which it is necessary to communicate a substantial amount of uncertainty, it can be helpful to have the assistance of a third party who "translates" the technical jargon into understandable terms and provides a respected voice to the community. Such a person must be seen not only as a responsible individual, but also as one who does not have a specific stake in the outcome of the communication effort.

The Role of Choice in the Perception and Acceptance of Risk

People and, collectively, communities ultimately seek to decide for themselves what risks they are willing to take. Experts who have studied the psychology of risk typically agree that people are much more accepting of risks that are of their own choosing—such as indulging in a high-fat diet or racing stock cars. This is because people are typically more willing to accept risks over which they believe they have some control. A good example of this is the person who is willing to do the driving on a busy highway, but who develops a severe case of white knuckles when riding as a passenger in a friend's (or spouse's) car.

Control over risk or perceived risk is frequently at the heart of conflicts involving environmental issues. This is why good community relations must be an inclusive process that encourages the involvement of stakeholders in the dialogue and that considers their input in the decision-making process.

In one case, residential neighbors of a contaminated site undergoing cleanup were willing to accept the preferred remedial alternative—on-site incineration of wastes—after the state agency responsible for the community relations effort suggested that an air monitor, which could be read by anyone who wished to, be installed in the neighborhood. This accommodation worked because one of the main issues of concern to the neighbors was their lack of control—their lack of ability to see for themselves that the mobile incinerator was not releasing emissions that exceeded air quality standards.

Although the air monitor would not *prevent* the release of excess emissions—and neighborhood residents were well aware of that fact—their ability to monitor air quality themselves provided a measure of control and comfort that made the use of incineration at the site acceptable to them.

Considering stakeholders' concerns or suggestions in the decision-making process can often result in solutions to environmental issues that are acceptable both to facility, site, or project managers and to the majority of community residents. Conversely, imposing a decision on a community without stakeholder input can result in hard feelings that may come out at some point in the future. Perhaps the community had no choice this time, but the next time a permit is sought or public input is required, the result may be quite different.

Risk Communication and the Media

The way the news media report risk can greatly influence individual perceptions. While the news media are often characterized in a negative fashion by persons who work in risk and environmental communication, when it comes to the reporting of technical information to the general public, they play an important role. Indeed, if there is a significant risk or an immediate danger associated with a certain situation, community relations specialists often work with the media to help get the information out.

Persons who are interested in obtaining detailed insights into the role of the media in risk communication should consult the work of Peter M. Sandman, who has written extensively on the topic of risk communication for environmental news sources. The following paragraphs reflect observations from the experience of the authors, other community relations specialists, and risk communication researchers on the role of the media in risk communication, particularly as it relates to community relations issues.

Scientists and technical experts who are charged with communicating about risk are typically the most critical of the media's role in risk communication. This is understandable, to some extent. By its nature, risk analysis requires a precise manner of speaking and thinking, while news reporting involves the provision of easily understandable information to the broadest possible audience.

The media, as well as many individuals, are not particularly interested in the intricacies of risk analysis, with all of its caveats and technical jargon. Instead, both reporters and community residents simply want the bottom line—Is the groundwater safe to drink? Is it safe to live next to this site? The fact that arriving at the bottom line usually comes at the conclusion of an arduous process of evaluating risk is, by and large, incidental to the story.

Additionally, in order for a news story to be more interesting and perhaps, even to be printed or broadcast at all, the potential for a higher level of risk is more likely to be emphasized over the potential for a low level of risk. The media are considered to have a duty to report all sides of a story, so reporters are naturally going to seek out persons with opposing viewpoints. If an "expert" can be found who disputes the government's or company's contention that a given level of risk is "safe," then that expert will likely be quoted in the story or presented in the broadcast.

The importance the media plays in forming risk perceptions should not be underestimated or taken lightly. Persons who communicate about risk must strive to provide reporters with simple, clear messages that concisely address the risk issue. This is the best way to ensure that the desired message conveying the true level of risk gets out.

PRINCIPLES OF CRISIS COMMUNICATION

As in the case of risk communication, crisis communication is a specialized discipline. Managers who will be involved in communicating during crisis situations will find helpful information in this section. However, thorough media training for those managers who will be acting as spokespersons, as well as assistance from professionals who work in crisis communication during actual crisis situations, is strongly recommended.

As the name implies, crisis communication is required in response to accidents, such as spills, explosions, fires, or incidents involving worker injuries or fatalities. Other situations requiring crisis communication may include the discovery of contamination at a site, impending enforcement actions, or the release of new data that suggest that a product or substance may pose a risk to human health or the environment. In some of these latter cases, community relations specialists will have some time to prepare for the disclosure of information to the public. In the former cases, immediate response is required.

It is best if positive relationships have already been formed in a community before a crisis; since community relations is a fairly new concept for many organizations, however, such relationships may not exist. Although a crisis is the least desirable way for the community to learn about a facility, site, or project, depending on the magnitude and the cause of the crisis, it can be possible to establish positive relationships afterward. Successful establishment of positive relationships after a crisis will hinge, in large part, on the public's perceptions of the crisis *and* on how it was handled.

The Crisis Communication Planning Process

Community relations specialists and managers of facilities, sites, or projects should take several steps to prepare for crisis situations. These steps include:

- Identifying potential crisis scenarios.
- Designating a spokesperson(s).
- Identifying the "crisis communication team" and other resources.
- Identifying persons who should be notified during a crisis.
- Assembling background materials and developing other materials for use in a crisis.

Identifying Potential Crisis Scenarios   The first step in crisis communication planning is to identify possible scenarios that could result in an accident, spill, release, fire, explosion, or injury. Often, this type of analysis is undertaken as part of a health and safety survey or emergency planning program for a facility, site, or project.

Community relations specialists frequently develop potential crisis scenarios with input from those who will be expected to respond when an incident occurs. Such scenarios can be used for drills or training exercises to help managers determine how they will handle calls—or the physical presence—of reporters and stakeholders and to develop general sets of message points and questions and answers that will help managers respond in a positive manner. Running through such scenarios—particularly with the assistance of a skilled media trainer—can help managers and community relations specialists identify the types of statements they want to make or avoid making, as well as the logistics for handling media and community inquiries.

Designating a Spokesperson A spokesperson and an alternate spokesperson should be designated during the planning process to ensure that all communication to the media and other stakeholders are funneled through them. This is necessary both to ensure consistency of messages and to ensure that important groups are not overlooked. It is extremely important that other employees or persons who are working at or involved with a facility, site, or project know that only the spokesperson is authorized to speak during a crisis, and that inquiries from the media or from other stakeholders, such as neighbors and elected officials, must be directed to this person.

For community relations purposes, the most appropriate spokesperson is, typically, the local facility manager or, in the case of sites, projects, or developments, the local project manager. Using this person as spokesperson reinforces his or her importance as the local community contact. Although community relations specialists and other professionals who specialize in crisis communication are often hard at work behind the scenes gathering information, coordinating the communication effort with senior management and legal personnel, and preparing messages and news releases, residents don't usually want to hear from communications folk—they want to hear from the man or woman in charge.

If the community relations specialist has been working in the community and has good contacts, then he or she can function as an auxiliary spokesperson with the media to provide background materials and coordinate the release of information. The community relations specialist can also maintain phone or face-to-face contact with elected officials, neighbors, and other concerned parties while the facility or project manager is dealing with the actual crisis.

Depending on the severity and duration of the crisis, senior managers from "headquarters" in either public or private sector organizations may also travel to the location to aid in coordinating the crisis response. In such cases, the local manager should usually maintain the communication lead in regard to local stakeholders and media (unless, of course, the situation is one in which the competence of the local manager has been called into question). Especially in the case of serious situations, however, the presence of a senior manager "from headquarters" to act as an alternate spokesperson and to meet with the media, elected officials, and others who may have been affected is important, since it communicates the organization's concern for the situation, its respect for the welfare and concerns of community residents, and its desire to ensure residents' safety and/or peace of mind.

Identifying the Crisis Communication Team and Other Resources
Responding to a crisis takes considerable effort, often under pressure and tight time frames. The spokesperson is usually busy managing the crisis as well as speaking to the media and elected officials. Thus, he or she will not have time to chase down information or write news releases (and facility or project managers don't usually have experience in writing news releases). A crisis communication team formed of support personnel should be assembled to assist the spokesperson.

Members of the crisis communication team should include a community relations specialist and, possibly, outside community relations or crisis communication consultants who can be brought in on short notice to assist, as well as a clerical staff member who can run errands, send faxes, and take messages. A technical support person should also be assigned to the crisis communication team to gather information of a technical nature. Last, the crisis communication team should include legal assistance to advise on the legal aspects of disclosing certain information. If an attorney is not available on-site, one should be available to participate in the team via telephone. This basic team can be expanded, depending on the magnitude and duration of the crisis.

Reporters typically look for a number of sources of information. Thus, it is wise to identify individuals who can be contacted to provide additional information during a crisis. These individuals may include toxicologists, remediation specialists, industrial hygienists, physicians, and others with the specialized expertise necessary to speak authoritatively on the potential crisis situations that have been identified.

Assembling Information and Developing Materials for Use in a Crisis
General information about the facility, site, or project should be identified for distribution as background material during a crisis. In the case of facilities, general information detailing the history of the facility, number

of employees, types of products and services provided (if the facility produces intermediate products, what the final products are), training that employees receive, and environmental safeguards, such as contingency planning, secondary containment, and pollution control devices may already be available. If such information is not available (as in the case of many sites or projects), general information providing the basic facts and descriptions should be developed.

Providing this type of basic information to reporters helps minimize inaccuracies in reporting, and, along with preliminary information about the crisis, can form the body of the first story about the incident.

Message points about the facility, site, or project, and even model news releases and general questions and answers, can also be prepared ahead of time. Availability of such "canned" language eases the process of preparing news releases and other information during an actual crisis.

Identifying Persons to be Notified During a Crisis   The planning process should consider what groups (media, elected officials, regulatory agencies, neighbors, and other stakeholders) need to be informed in the event of an incident. Phone lists of key individuals, such as elected officials and neighbors, as well as media contacts, should be developed—both for crisis communication purposes and for follow-up after the incident. If a facility, site, or project has a community relations program, then such groups and individuals have probably already been identified.

A Word About Working With Attorneys— and a Word to Attorneys

Since crisis situations can lead to lawsuits and enforcement actions, it is important to work with attorneys to ensure that the information that is released will not create additional problems. This said, attorneys need to understand the importance of releasing information to the media and the public. Attorneys' primary focus is on protecting the interests of their clients, and their first reaction will typically be to restrict the flow of information.

The authors agree that all crisis communication must be carefully thought out, preferably with the assistance of an attorney. We suggest, however, that some attorneys do not always realize how placing a stranglehold on the release of information can seriously damage an organization's relationship with its neighbors, with serious financial and emotional consequences. Persons working on crafting information in response to crisis situations should work with the organization's attorneys, but the attorneys also need to listen to those who specialize in communication and community relations.

Understanding the Media's Role in Crisis Communication

Persons who do not regularly work with the media, or whose only brush with reporters is during crisis situations, often don't understand how reporters function or what they are looking for. Reporters are looking for a story. That is their job. And crisis situations are stories. If you don't talk to them, they will find someone who will—and that someone may make statements that are extremely damaging, especially when their comments are coupled with the statement, "Facility representatives were unavailable for comment." Contrary to the perceptions of many, reporters are rarely "out to get" anyone. Most respond well when they are provided with information and with some respect.

Managers need to remember that the media perform an important service. The media inform the public about current events and can convey important information. You can affect the accuracy of the media's reporting by providing factual and timely information to reporters. News releases about crisis situations and background information on the facility, site, or project are often picked up verbatim if they are well written and objective.

Working with Reporters Reporters work on deadline. When a crisis occurs and reporters appear at a facility or site or call on the phone, find out what their deadlines are and get back to them with whatever *confirmed* information you have available beforehand. Although one strategy in a community relations program is to develop relationships with one or more local reporters—which can certainly enhance their ability to report accurately about a crisis at your facility, site, or project—don't play favorites in the release of information about the crisis. Release information to all reporters at the same time.

Treat reporters as professionals. The authors also believe that expressing appreciation to reporters for their role in informing the public is a good idea. Treating reporters as adversaries is *not* going to improve the situation. Although news reporting is "objective," positive or negative attitudes on the part of the reporter can still be reflected in the wording of the story and the order in which facts are presented, both of which affect the "tone" of the story and public perceptions.

Most reporters have little knowledge of the technical and regulatory aspects of environmental issues. Thus, information of a technical or regulatory nature has to be carefully spelled out for them so they can report accurately. This can place quite a burden on the manager who is serving as spokesperson, which is one of the reasons why, in crisis situations, the team approach, with specialists in community relations and crisis communication laboring behind the scenes preparing information for release, is so important. Reporters' expertise—and it definitely is expertise—is in gathering together information from diverse sources and assembling it quickly into a coherent story. This is a daunting task, given that reporters are thrown into situations about which they may know little. In such cases, their reporting can only be as good as the information that *you* give them.

Reporters will frequently ask hypothetical questions or ask a spokesperson to speculate as to the cause of an accident, monetary value of the damage, and so forth. This line of questioning is part of the usual process reporters use to attempt to draw out information and flesh out the story. Spokespersons must resist answering either type of question, and point out that these questions cannot be answered because they are "hypothetical" or "speculative."

Three other mistakes that spokespersons can make when dealing with reporters are continuing to talk after the point has been made, saying "no comment," or going "off the record."

Don't let silence entice you into saying more than you are prepared to say—invariably, rambling on will lead to speculative statements or offhand remarks that aren't well explained and that can be misunderstood. The statement "no comment" is frequently perceived by the public as an admission of guilt. Within the perspective of community relations, "no comment" is irritatingly smug and superior, translating into, "I don't have to tell *you*." If you aren't ready to release information, or don't know something, then an appropriate answer would be, "I don't have that information at this time" or "We don't yet know what the cause was; we are investigating now" or something else that communicates a desire to be open, even if there isn't any confirmed information yet.

In regard to going "off the record," don't do it. Nothing is ever "off the record" with reporters. Not only do reporters generally consider anything you tell them fair game, but they may take an "off the record" comment and use it as a point of departure for interviewing other sources (e.g., "He said it could have been the boiler, but they haven't confirmed it. What do you think?"). Additionally, if you allow "off the record" comments to slip into your conversation, a reporter working on deadline, no matter how friendly, may simply forget what is or isn't to be released.

Additional Crisis Communication Concepts

Three other important concepts should be considered in crisis communication from the community relations perspective. The first of these is that you can accept responsibility without necessarily taking blame. People want to know that if something happens in connection with your facility, site, or project, you will act to safeguard human health and the environment.

For example, if a supplier's tanker truck spills a chemical at your facility, community residents' first concern is that the spill be cleaned up and that risk to residents and the environment be minimized. Thus, the facility spokesperson should be addressing the problem—not who is at fault. He or she should stress that since it is his or her facility, the number-one concern is cleaning up the spill and minimizing risk to human health or the environment. You can say straight out in response to a question about who is at fault, "We will look at that later. Our number-one concern is cleaning up the spill and protecting the environment." Casting blame elsewhere, even if justified, looks evasive and doesn't address the key issue—containing or cleaning up the accident.

Second, you can acknowledge public concern without confirming its validity. In other words, you can say that you know that people are worried about the spill. Follow this acknowledgment with information about the measures being taken to clean it up and, possibly, the progress you expect to make. If applicable, provide health or environmental impact information about the chemical in question as such information becomes available—but don't omit acknowledging the public's concerns. People do not want to think that their fears are being dismissed out of hand.

Last, and, in many instances, most important, if someone has been injured or killed, or the environment has been damaged, it is *always* appropriate to express sorrow and compassion, regardless of how or why it happened. This is just human decency. If you were to hear of such an accident elsewhere, wouldn't you express compassion? If fault for the accident lies with the organization responsible for a facility, site, or project, failing to express compassion is not going to prevent lawsuits or the public's disapproval, but it will make it that much harder to regain the community's support or trust. Some of the bitterest words uttered by community residents are, "They didn't even say they were sorry."

Public Perceptions of a Crisis

Both the reporting of crisis situations and public attitudes will reflect two factors. The first is the crisis itself. Obviously, crises that occur because of negligence or that cause serious injuries or fatalities or significant damage to the environment and to wildlife will be viewed in a very negative light. Organizations that are responsible for facilities, sites, or projects that cause or are the scenes of such accidents will be subject to negative news stories and community anger. Crises of lesser magnitude and those that do not involve gross negligence are typically easier to overcome.

Of course, from a community relations standpoint, other community concerns or issues and prior attitudes toward the facility, site, project, or organization responsible for it (or toward similar facilities, sites, projects, or organizations) will also have an important effect on how the crisis is viewed. For example, if a facility is already disliked by residents because of labor strife or other damage (or perceived damage) to the environment, even a minor incident may be enough to mobilize major anger and opposition from community residents.

The second factor that affects both reporting and public attitudes has to do with how the organization responsible for the facility, site, or project responds to the crisis. Did it try to blame others? Fail to reach out to injured parties or to mitigate environmental damage? Resist providing information? Lie or make misleading statements? Resist working with emergency responders or other stakeholders to respond to the crisis?

The way an organization handles a crisis is frequently the topic of many of the news stories that are published about the event, and if the organization has done a bad job—including in the release of information about the crisis—those stories will be negative. In regard to community relations, the loss of trust on the part of stakeholders can be devastating. People may be able to forgive an accident, but they find it a lot harder to forgive poor handling of that accident. After all, substantial effort in community relations is expended on providing stakeholders with a level of comfort regarding the competence of the people in charge of a facility, site, or project. Fumbling the handling of a crisis—including the communication—destroys both that level of comfort and the trust that community residents may have placed in the managers.

Model for Immediate Response to the Media

A comprehensive communication response to a crisis typically requires the assistance of people who are well versed in media relations and crisis communication. However, the designated spokesperson is often on his or her own when a crisis first occurs. A standardized approach used by the authors and other community relations and media specialists can be used to successfully handle such "first contacts." This model sets a positive tone for contact with reporters and provides some reportable information in preparation for either the arrival of media relations/crisis communication assistance or the preparation of other *confirmed* information. This model can be used within the first few minutes, in the absence of a prepared statement:

Step 1. Confirm the incident (fire, accident, spill).

Step 2. Say that the specifics are under investigation, and, if applicable, that actions are being taken to contain and control the situation. Say that you are part of the investigation and/or the efforts to contain or control the situation and have to return—but you wanted the reporters to know that you are the media contact. If you have general background information on the facility, site, or project, it should be provided to the reporters at this time.

Step 3. Ask the reporters for their names, phone numbers, newspaper or broadcast media name or call letters, and deadlines and say that you will get back to them with information. (If no confirmed information is available by deadline, get back to them anyway, reiterating that confirmed information will be released as it is available).

Step 4. Thank the reporters for helping inform the public, and reiterate that, as information is confirmed, you will make it available.

Step 5. Walk away. Reporters can't very well fault someone who is trying to manage a crisis for leaving hastily after a few remarks.

Dealing with the Aftermath of a Crisis

Organizations that handle a crisis well, including providing information to the media and to other key stakeholders, such as local officials, as soon as it is confirmed, can often recover from incidents that aren't results of gross negligence. Organizations with positive resolutions to crises are those that have responded quickly and decisively to the incidents themselves. Their responses to questions are also clear and concise, and they take the time to ensure that the information is understood. Most importantly, organizations that successfully weather crises tell the truth and accept responsibility for what has happened.

Once the crisis is over, additional work will be necessary to re-establish trust and good relationships (or establish trust and relationships, if no community relations work was done beforehand). This effort typically entails one-on-one and small group meetings with formal and informal leaders and neighbors, follow-up press releases on the findings of investigations, and mailings to certain stakeholders describing the actions that were taken, investigative findings, or other such information. We mentioned earlier in this section that in the case of serious accidents, senior managers sometimes come to the site to lend their assistance. Bringing in "headquarters" people for a round of follow-up visits to neighbors, elected officials, and other stakeholders can send a powerful message about the organization's desire to work with the community.

Expect that the facility, site, or project will be known as "the facility, site, project, where the fire/spill/explosion/etc. happened" for a long time. You should also expect that, if a similar accident happens in your area (or nationally, depending on the magnitude of the original incident), your facility, site, or project will be back in the news. If formal questions and answers or other information about the incident were not prepared during the crisis, prepare them afterward so you can continue to answer the inevitable questions in a consistent and positive manner.

Disclosing Information to Mitigate an Unfavorable Response

Although most crisis communication involves a sudden event, such as a chemical spill, occasionally, managers will get wind of a problem before it is upon them—and before either the media or the neighbors are at the door asking questions. As we noted earlier in this section, such situations can involve the discovery of contamination, impending enforcement action, or new data on the risk posed by a product or process.

In all of these cases, it is likely that information about the situation will eventually come out. Thus, it behooves the organization to release it first, both to increase the amount of control it will have about the initial messages and to maintain a proactive stance in the community. Although disclosures of this nature must be carefully timed and carefully planned, organizations that release information in a proactive manner typically score points for being up-front. Such disclosures should include information on the measures that are planned or that will be taken to remedy or address the situation.

As we said in chapter 1, the public frequently views omission as a form of lying, and failing to make a disclosure about a serious issue, especially if it is perceived as having the potential to affect human health and the environment, could be taken as an omission and a lie. Thus, failing to disclose information that is likely to come out anyway not only means a loss of control over the way the messages are presented, but it also can lead to serious distrust on the part of the community. Conversely, well-executed disclosure can enhance an organization's credibility in the community.

In one case, the management of a facility discovered that contamination from the facility property had migrated off-site. It didn't matter that the contamination had been caused by previous owners of the facility—an investigation and cleanup on neighbors' property would have to be conducted. To make matters worse, the contamination was PCBs, definitely a "hot button" in the community. Fortunately, management was already engaged in a dialogue with its neighbors and elected officials about the facility's operations.

The facility's management elected to disclose the information to the community rather than allow residents to find out about it secondhand—an inevitability, since the environmental agency had already been notified. Managers met with the neighbors, elected officials, and others who had expressed interest in or concern about the facility, and they released the information to the media. Although no one was pleased about the contamination, the overall response from the community was still positive. Stakeholders appreciated being told and they appreciated the fact that the facility's management had laid out a plan to fix the problem. This response was mirrored in the media—the local newspaper ran an editorial praising the facility's management for its handling of the situation. The facility management took the "crisis" out of the communication through a well-reasoned and proactive approach.

CHAPTER

4

COMMUNITY RELATIONS ACTIVITIES

This chapter covers some of the more commonly used activities for communicating with the public and involving interested parties in a community relations dialogue. Specific community relations activities may be especially suitable for providing information and soliciting input, building and maintaining relationships with stakeholders, or fulfilling both of these functions. Most community relations programs will involve the use of a mix of activities to suit the issue and the stakeholders who have an interest in the issue.

This chapter looks at "neighbor" and stakeholder visits, the linchpin of most community relations programs; events, such as public hearings and tours; written materials; considerations for working with the news media; and other "maintenance activities" designed primarily for relationship building. Additionally, we explore community advisory panels (CAPs), vehicles that are increasingly used to gain input from stakeholders and educate representatives of stakeholders groups about the facility, site, or other issue in question.

DEALING WITH THE RELUCTANT
AND THE NOT-AT-ALL RELUCTANT

Community relations programs are usually marked by "feast or famine" when it comes to public participation in the dialogue surrounding an environmental issue. The "feast" scenario typically involves stakeholders who turn out in droves and are loudly opposed to or demanding regarding the issue in question. Such situations pose their own challenges, but getting people involved generally isn't one of them—although getting them into a productive dialogue frequently is.

The "famine" scenario involves trying to conduct a dialogue—often to fulfill statutory requirements or when attempting to establish a relationship with the community *before* an issue that could create major concerns emerges—amid the deafening silence of apparent disinterest. These situations can be as unsettling for community relations specialists as managing angry community residents, especially if the community assessment has indicated that concerns or the potential for concerns are indeed present. Thus, the challenge of community relations activities is to entice people into a dialogue *before* their questions or concerns erupt into major controversy, or, conversely, to work productively with stakeholders who are already forcefully demanding a say in the decision-making process.

Classifying Community Relations Activities

Community relations activities are geared, to a greater or lesser extent, to communication and to relationship building or maintenance, as shown in Figure 4.1. Ideally, a community relations program should include activities

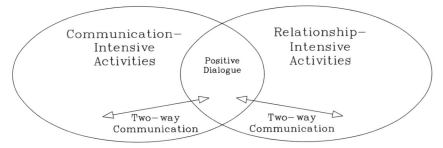

■ **FIGURE 4.1 Both Communication-Intensive Activities and Relationship-Intensive Activities Are Often Needed to Promote a Positive Dialogue.**

that will do both. Community relations activities can also be categorized in terms of their use in "issue-specific" situations (e.g., a public meeting used to present information on a RCRA permit) or as "maintenance" activities (e.g., talking to the neighbors on a regular basis simply to maintain contact and keep channels of communication open in the absence of significant concern). A number of the activities that we discuss in this chapter could be used either as "issue-specific" or "maintenance" activities. How each activity is used will determine what its primary objective and relative importance should be within the community relations program.

Two of the considerations that must be made in designing a community relations program are the manner and the extent to which stakeholders wish to be involved. Although some groups may desire only the occasional update to mark significant events, others may desire an ongoing dialogue, tours, and significant amounts of information. Thus, a number of activities may be undertaken to meet the wants and needs of these different groups.

Different Goals Require Different Approaches Attempting to meet different goals for different types of issues may require a different mix of activities—even in the same community. For example, community relations activities in support of the cleanup of contaminated sites or decision-making processes regarding development or use of natural resources are typically dialogue-intensive. Thus, the "mix" in regard to community relations activities will necessarily be communication-intensive.

Community relations programs in support of facilities and developments that make long-term use of natural resources, such as quarries, should also include communication-intensive activities, as well; however, they are likely to include more "soft," relationship-building activities that help the organization in question establish durable, positive relationships in the community and promote images of good organizational citizenship.

Additionally, although some "maintenance" activities—such as sponsoring science projects at a local school—promote a dialogue indirectly, such activities can help promote a familiarity and a level of comfort that can ease the process of building relationships so that a dialogue on more serious issues can be established.

VISITS TO NEIGHBORS AND OTHER STAKEHOLDERS

The linchpins of most effective community relations programs are informal visits or briefings to neighbors and other primary stakeholders. These activities typically involve one-on-one (or very small group), face-to-face meetings with neighbors and other primary stakeholders, such as local officials and environmental or social group representives. The immediate and informal nature of such meetings can promote the frank exchange of information and the establishment of bases for positive personal relationships. Such meetings are often the heart of community relations, and no amount of written material or public meetings can substitute for these types of exchanges.

The following paragraphs concentrate on neighbors in particular; however, considerations in regard to neighbors also track closely with those regarding visiting and briefing other stakeholders.

Neighbor Visits

Visits with neighbors can be one of the most important and effective tools for promoting a dialogue with the stakeholders who are often the most intimately affected by an environmental issue. When we think of neighbor visits, residential neighbors most often come to mind; however, managers should also remember that other businesses and landowners, both public and private (including wildlife refuges, vacant property, parks, and the like) can be neighbors, and they should also be contacted.

In the case of contaminated sites, informal visits with neighbors by Environmental Protection Agency personnel or businesses engaged in voluntary cleanups of their property can often encourage neighbors to ask questions and voice their concerns much more readily than meetings in formal settings. Sitting down with neighbors in their own homes conveys respect and interest in their views and concern for their peace of mind that cannot be duplicated through other means.

Neighbor visits are especially important in cases in which investigative or remediation work must take place on or directly adjacent to a neighbor's property. The informal chat that provides frank answers and direct information on what needs to be done—and why—can often secure willing cooperation and access to property that otherwise could require delays and potentially costly legal action. Face-to-face meetings allow the person responsible for community relations to work out details that could lead to opposition if they aren't discussed. For example, a neighbor may be concerned about the appearance of a monitoring well in her yard; providing the neighbor with the option of having a flush-mounted well (a well installed so it terminates flush with the ground rather than having a casing that sticks up out of the ground) can easily overcome this opposition. Regular communication, via telephone or in person, regarding the progress of the investigation or cleanup should take place throughout the duration of the activities.

Effective neighbor visits can also repair strained relationships that can develop during investigation and cleanup activities. In one case, a community relations specialist was asked to work with neighbors of a site who had had their yards dug up as part of the remediation process. Additional soil removal was going to be necessary in the residents' yards—activity to which they were adamantly opposed. The specialist's interviews with the neighbors—and her own visual examination of their properties—revealed that, at this point, they were less concerned about the contamination than they were with the condition of their properties.

The neighbors finally agreed to allow the remediation to continue after the specialist worked with them to secure an agreement with the contractor to restore their properties to their previous condition after the cleanup was complete. As part of the agreement, the specialist photographed each neighbor's property to provide benchmarks for restoration. Each neighbor "signed off" on the photographs, providing them—as well as the contractor—with an objective measure for the restoration.

Fact sheets and other written materials can be given to neighbors to provide them with additional information about current or impending activities. Persons conducting neighbor visits should also always leave a business card so that the neighbor can contact them if they have additional questions or comments.

Community relations programs in support of facilities, operations involving the harvesting or extraction of natural resources, or major public works or transportation-related projects should also include neighbor visits. As in the case of neighbor visits in support of site investigations and cleanups, the initial contact should ideally be in person, although follow-up conversations can often be handled over the telephone.

There are several circumstances under which neighbor visits in conjunction with facility community relations efforts may be conducted. Neighbor visits are advisable prior to the commencement of permitting activities or the expansion or alteration of a facility, so that neighbors' questions or concerns can be addressed prior to a public hearing or public comment period. As we will discuss in chapter 6, opposition to a facility, site, or project often arises as a result of misinformation or lack of information. Answering questions and addressing concerns in an informal setting can often nip such opposition in the bud. Additionally, receiving input regarding neighbors' preferences may lead to accommodations—such as construction of berms to block noise—that can minimize opposition and lead to better decision making.

Neighbor visits by facility or project personnel also often occur in response to a complaint or an accident that has had off-site consequences or created off-site concerns. Such visits are extremely important, since ignoring neighbors in such circumstances can set the stage for serious opposition and ill feeling down the line.

Neighbor visits in the aftermath of an accident or in response to a complaint often do more than any other action a manager could take to establish a positive dialogue with the neighbors. If the neighbors are angry, the opportunity to express their anger in person can help defuse the emotion, provided that the organization's representative does not react in a defensive manner and problems that need fixing are corrected. Insincerity or untruthfulness on the part of organization representatives will invariably compound whatever problem the neighbors are concerned about and make it that much harder to establish a positive relationship in the future.

Neighbor visits are not a panacea. Long-standing problems, such as repeated spills or discharges onto a neighbor's property or evidence of repeated violations of requirements for odor control or emissions, cannot be charmed away by neighbor visits. Under such circumstances, neighbors may refuse to meet with facility personnel whom they perceive as acting in bad faith. Neighbors may also refuse to meet with organization representatives out of fear, or because they believe that such conversations may jeopardize their opportunity to sue for damages. In such cases, it is better not to try to push contact. Instead, meet with neighbors who are willing to talk and demonstrate a commitment to environmental protection by taking the actions necessary to correct problems.

Neighbor Visits as "Maintenance Activities" Neighbor visits may also be conducted on a routine basis to ensure that the lines of communication remain open. For example, a facility manager may make it a point to drop by or make telephone calls to facility neighbors once a year or so. If the neighborhood is large, the visit or call may be made to the head of the local homeowners' association. Although this type of contact may seem insignificant—frequently, if the neighbor has no issues, this yearly conversation is only a few minutes long—it supplies the neighbor with a name and a face or voice and greatly boosts the probability that he or she will contact the facility representative if a question or concern arises. Such contacts also make it easier to initiate a dialogue with the neighbors in the event of a planned expansion, permitting activities, or accident. Routine contacts help minimize accusations such as: "I never heard from them until there was a problem," or "I never heard from them until they wanted me to agree to their zoning variation request."

Visits to Other Stakeholders

The concepts that we have just discussed in conjunction with neighbors also apply to visits to other stakeholders. Keeping the lines of communication open with primary stakeholders, particularly through face-to-face contact, can build the relationships necessary to get through some very tough problems. As in the case of neighbors, visits can also be conducted via telephone after initial face-to-face meetings. Telephone contact may be preferred over face-to-face meetings by some busy officials and other stakeholders—especially those who do not live or work near the facility, site, or project. As in the case of neighbors, "maintenance" visits or phone calls should also be made on a regular basis to other important stakeholders.

WRITTEN MATERIALS

Written materials constitute one-way communication and should not be used as a substitute for actual conversations with primary stakeholders. Written materials do provide important supplements that can add substantially to the quality of the dialogue, however. Written materials convey information necessary to ensure an informed dialogue by providing stakeholders with information they can refer to, show to others, and use as a basis for conversing on topics that they might otherwise know little about. **All written materials used for community relations purposes should include a contact name and phone number and an address for making written comments or inquiries.**

Written materials are part of the mix that should be used in most community relations efforts. In this age of busy schedules, a simple fact sheet or newsletter can provide an update to people who may be too busy to attend a meeting, but who are interested in keeping up with what is happening in regard to a facility, site, or project. The balance between paper and talk should be carefully weighed, with attention paid to stakeholders' preferences regarding the amount and format of materials. Although it is wise to prepare information on all major topics that may be of interest to stakeholders, community relations specialists should not appear to be shoving paper at residents in place of speaking with them. Still, the availability of well-prepared written materials can help convey a willingness to provide information to the public and help minimize accusations that persons in charge of a

---

The "Slickness" Factor

One variable that must always be considered in regard to written materials is the impression created by their overall appearance—and whether this impression is in keeping with the intent of the community relations program. Community relations specialists must balance the need to create materials that are clean, attractive, and easy to read against an appearance that will be perceived as "slick." The line between the two will vary from community to community. While a four-color brochure describing a proposed plant expansion may be well accepted in an upscale community (and, in fact, a publication of lesser production values may be viewed either with scorn or annoyance that the facility in question didn't think enough of its neighbors to "spend some money on something decent"), the same brochure may intimidate and anger residents of more modest means in another community, who may perceive the publication as evidence that the company is willing and able to spend a lot of money to get what it wants.

Corporations, in particular, may have trouble producing materials that don't look "slick" to some neighbors. Most companies have standards regarding the quality of sales materials and business correspondence that they may, without thinking, transfer to the preparation of materials for community relations purposes.

While community residents are likely to accept a glossy, professionally done annual report for a *Fortune* 500 company, since annual reports for big companies are *supposed* to look expensive, a hand written note from a facility manager to a neighbor thanking him or her for participating in a focus group will usually make a more favorable impression than a neatly word-processed letter. Thus, appropriate production values will vary depending on the type and intent of written materials, the community, and the receiver of the information. Businesses shouldn't automatically apply the level of formality that they might use with a customer or supplier to materials intended for use in environmental community relations.

facility, site, or project are changing their stories to suit a particular audience.

Whenever possible, drafts of written materials should be previewed by focus groups or other readers who are not overly knowledgeable about technical and regulatory issues *before* they are distributed, to ensure that the materials are understandable and that they make the right impression. Additionally, efforts should be made to follow up with certain primary and secondary stakeholders and others who received the materials to gauge the materials' effectiveness and to see if there are additional questions regarding the subject matter that have been triggered by the information. Requests for additional information should not be considered a failure if they reflect interest in learning more about the subject. If they reflect lack of understanding (e.g., the materials were too complex) or distrust for the information thus far given, then the community relations specialist will have to modify his or her approach.

Consistency Among Written Materials

The number of written pieces that may support a community relations program will depend on the issue and the subject of the community relations program (facility, contaminated site, short-term project, and so on). Consistency of both terminology and facts is extremely important. In both cases, terminology must be defined, if it is unfamiliar, and facts that are contingent on certain conditions must include discussion of those conditions.

In one case, a profile about a facility stated that it had never had a "reportable spill." Another section of the profile discussed the fact that the facility had secondary containment, which prevented the small spills that take place every so often from reaching the soil. A local reporter who read the profile decided to write an exposé on "unreported spills" at the facility, and got quotes from a former employee who said that, while he was working there, several spills had occurred.

Both statements in the facility profile *were* correct. But since the first statement had failed to define the regulatory definition of a "reportable quantity," the reporter, and, unfortunately, a number of community residents, believed that the discrepancy was evidence that the facility's management was lying about its spill history.

Fact Sheets

Fact sheets comprise the backbone of written community relations materials about environmental issues. Fact sheets are short informational pieces designed to explain regulatory or technical issues, to provide overviews or updates of activities related to a specific issue (such as a cleanup, permitting process, or response to an emergency situation), or to provide background information or an overview of a specific topic. Since fact sheets are typically intended to provide information to persons, such as community residents, who aren't well acquainted with the technical and regulatory aspects of environmental issues, readability is very important. Use of easy-to-read diagrams, charts, and other graphics that add clarity is always desirable.

Other techniques that can enhance the readability of fact sheets include the use of headings; boldfacing of technical terms and providing easy-to-understand definitions within the text and/or in a glossary block; and the use of a question-and-answer format that leads the reader through key issues and concerns. Short paragraphs, short sentences, and simple words all enhance readability.

Fact sheets are typically short. One to four pages is a good length, although some that are providing an update on the cleanup of a contaminated site, for example, may be as long as eight or 10 pages—including a number of diagrams, charts, and other graphics. In cases of situations involving several issues or complex, multifaceted issues, it is usually better to produce several short fact sheets than a single, long fact sheet. Using several short fact sheets is also useful in cases in which certain stakeholders may be more interested in some issues than in others, since using individual fact sheets, rather than "burying" some issues in a longer, single fact sheet, better conveys the message that each issue is important.

In the following example, one important stakeholder group was especially concerned about increased truck traffic in connection with a facility. By producing a series of short fact sheets, this issue, which was of paramount importance to these stakeholders, could be given adequate attention without "cluttering up" a longer fact sheet with information that wasn't of interest to most of the other stakeholder groups.

A facility that was planning to expand and to modify several permits to accommodate increased operations elected to produce a packet of four fact sheets—each one page, double-sided—to describe the expansion and permit modifications. These fact sheets provided overviews of the following topics:

■ Facility history and description of operations, including a discussion of why the expansion was needed and what it would entail.

- The actual building expansion, including a sketch of the current and proposed facility plans.
- Permitting and environmental protection, including a description of the process of modifying RCRA and wastewater discharge permits and discussion of enhancements to the facility's wastewater treatment system and its pollution prevention/waste minimization program.
- Transportation safety, including descriptions of routes that trucks traveling to and from the facility would take, the expected increase in truck traffic that would occur as a result of the expansion, hours of operation, safety precautions drivers are instructed to take, and traffic data from the state department of transportation.

Newsletters

Newsletters may focus on activities at a site, facility, or project. Newsletters differ from fact sheets by providing general news coupled with short articles on a variety of different topics. Newsletters may serve to tie other communications on a site, facility, or project together by acting as an easy-to-read overview of activities that may have otherwise been documented by the more detailed fact sheets, workshops, or public meetings. Newsletters may also be used to reach secondary stakeholders who are not actively participating in the dialogue, but who should be kept apprised of major developments. For example, a newsletter on a large site might provide an overview of activities that have occurred over a set period of time; report on stakeholder involvement activities, such as community advisory panels, technical workshops, or meetings with senior EPA officials; and advertise the availability of fact sheets or the dates, times, and locations of upcoming meetings.

Newsletters may also be used by public or private sector organizations or facilities to provide updates of general information. Some of this information may be directly linked to environmental issues, but other "articles" frequently cover issues such as personnel; impending developments, such as plans for expansion or upgrades; and status of activities (in the case of government agencies, the number of complaints investigated, permits issued, and so on; in the case of facilities, the number of units produced, gross revenues, and the like). This type of general information promotes some understanding of what a facility, company, or government agency does, conveys the message that inquiries are welcome, and provides contact information. As in the case of the more targeted newsletters used for some large or high-profile contaminated sites, general newsletters are a good way of contacting secondary stakeholders.

Questions and Answers

Questions and answers, as we discuss them here, are not typically distributed to the public. We include them here because they are important in-house written documents to support the community relations effort. As discussed in the section on fact sheets, the question-and-answer format can be extremely useful in fact sheets, since it leads the reader easily through a logical sequence of information; however, questions and answers are also often developed for use by managers who have to provide verbal information to the public, whether through public meetings, neighbor visits, or media briefings.

The development and use of questions and answers enhances consistency in facts and terminology. Additionally, by preparing questions and answers and working with them ahead of time, good, understandable answers can be developed for sensitive questions or to explain technical subjects in language that is simple, yet accurate. Technical personnel often have trouble figuring out how to explain complex concepts to laypersons. Development of questions and answers can help them pick through the minefield of technical terms so they can provide answers that are readily understandable.

The process of preparing questions and answers helps community relations specialists identify key issues and determine how best to discuss them with stakeholders. As in the case of other information developed to address a specific site, facility, or project, questions and answers should be tailored to the community in terms of level of detail and topics of concern.

Questions and answers for verbal use are not intended to be read verbatim. Instead, they represent a mechanism that allows persons who are charged with communicating with stakeholders to "try out" various answers and explanations so that they can determine beforehand how best to provide answers that are complete, accurate, and easy to understand. Especially in the case of technical and regulatory issues, this is not easy. By tackling questions on paper first, community relations specialists, plant or project managers, or others who will be speaking to the public can determine what information they need to convey to make an answer understandable.

For example, a project manager for a contaminated site may need to discuss how samples are being collected and analyzed. Invariably, some

WRITTEN MATERIALS ■ **131**

technical terms will have to be defined to ensure that listeners can understand what is said. By looking at proposed answers ahead of time, community relations specialists can integrate the necessary definitions into the answers and minimize confusion.

Criteria for Questions and Answers The preparation of questions and answers involves identifying as many questions that stakeholders have or are likely to ask as possible and then developing answers that are direct, truthful and complete, consistent, and easy to understand. In the case of technical issues, this can be extremely demanding. Let's look at the elements we've just discussed:

- **Direct**—answers must answer the questions. Although they may include detailed explanations *and* convey uncertainty, anyone hearing an answer must be able to come away with an understanding that the speaker said "Yes," "No," "We don't know yet, but we are continuing to investigate . . . ," "We intend to, but the timetable will depend on the weather . . . ," or some other definitive statement.
- **Truthful and complete**—answers must not appear to be shying away from the truth or omitting important elements that could be construed as constituting a lie, such as stating that contamination is present in the soil—but not mentioning unless asked directly that it has also been detected in the groundwater.
- **Consistent**—answers should be consistent with written information, such as fact sheets, in terms of terminology and the information they convey. One of the reasons why development of formal questions and answers is so desirable is to enhance consistency in the messages that everyone attached to a facility, site, or project communicate to the public.
- **Easy to understand**—answers must be readily understandable if they are to do what they are supposed to do—answer people's questions. Technical and regulatory terminology should be explained. Comparisons and descriptions should be used as much as possible to help listeners visualize what they are being told.

After questions and answers have been prepared, they should be read by as many people as possible to determine whether they meet the criteria listed above. If possible, the questions and answers should be read by at least one person who is not knowledgeable about the issue, to see if the messages they receive are the ones the answers were intended to convey. Attention should also be paid to readers' reactions to the tone of answers. Answers should be neither abrupt nor give the impression that the speaker is trying to skirt an issue.

Question and answer development should begin with an analysis of the community assessment. Many of the major questions, along with an indication of the type of information stakeholders are looking for, are readily identifiable in the community assessment. Once these core questions and answers are identified, additional, follow-up questions and answers that are natural extensions to the stakeholders' learning process can be easily identified.

Message Points

Message points (or "talking" points) are lists of the key concepts or messages that managers wish to communicate to stakeholders regarding an issue. As in the case of questions and answers, developing message points before engaging in conversations with stakeholders is advisable.

Message points should also meet the criteria we just discussed for answers—they should be direct, truthful and complete, consistent, and easy to understand. However, message points should also be brief—preferably one or two sentences (public relations practitioners often refer to such messages as "sound bites"). Thus, message points are definitive, summary statements of key concepts. Additional explanation, as necessary, is given after the message point. Message points are often developed in conjunction with questions and answers.

Issue Statements

Issue statements can be likened to editorials or position papers. They provide statements of policy or opinion about specific issues and explanations of those policies or opinions. For example, an issue statement may be prepared to explain why a company supports a particular environmental regulation or program—or why the company finds it unnecessarily burdensome. Issue statements may also be used by government agencies to explain their stances on particular issues.

Issue statements are most likely to be used in conjunction with fact sheets, questions and answers, and other face-to-face community relations tools in complex situations in which questions are being raised about why an action is—or isn't—being taken. As such, an issue statement provides an explanation that goes beyond the technical or regulatory issues of the situation at hand to discuss the reasoning behind a policy decision.

For example, a fact sheet may describe the use of a toxic chemical at a facility and the handling procedures that are used to ensure that it is managed safely. The fact sheet may also include information on the reason why the chemical is necessary to manufacture certain products.

Since fact sheets are typically represented as neutral, informational documents, the organization may also wish to put together a supplementary piece—an issue statement—that provides additional information on the organization's policy toward the use of such chemicals, specific reasons why the organization chooses to use the chemicals, and its policy toward managing risk and ensuring the health and safety of its employees and neighbors. Although much of the information would be the same as that used in the fact sheet, the emphasis on organization policy, with the implicit declaration of organizational responsibility and judgment, sets the issue statement apart as a unique type of document.

Thus, issue statements are used to emphasize an organization's reasoning and positions with explicit statements about important issues. Issue statements may range from half a page to three or four pages, double spaced. Longer documents are typically considered position papers. A well-written issue statement can provide focus for a community relations dialogue, since it provides insight into the overall reasoning and direction that is guiding a public or private sector organization.

Brochures

Brochures are typically developed to provide either general overview information, as in the case of brochures on companies or facilities describing what they manufacture, or in-depth information on a particular topic. For example, a company that manages wastes may develop a brochure that describes how waste is generated, the regulations that govern its management, and the facilities and procedures that the company uses to ensure its safe treatment, storage, or disposal. Such brochures are bigger, fancier versions of fact sheets.

In community relations, brochures are usually used as supplementary materials, since they are rarely specific to the community. Instead, they can serve as backup materials that can be used as part of a mix of written materials for providing information to stakeholders.

Environmental Annual Reports

Environmental annual reports are a relatively new type of communication tool. It should be noted that not all "environmental annual reports" are published yearly. Some are published on a two- or three-year cycle. We use the term environmental annual report here to differentiate these reports, which are communication tools designed to meet public needs, from other types of reports that are prepared to fulfill regulatory or technical requirements.

The goal of the environmental annual report is to provide specific information on environmental performance, such as the amount of air emissions, wastewater discharges, or spills, as well as pollution prevention and waste minimization efforts. This information, which ideally can be compared from year to year, is intended to allow stakeholders to follow the progress of the company's improvements in environmental performance over time. Much of the "hard" data in these reports for manufacturing companies come from the Toxics Release Inventory (TRI), which is publicly available. (Note: utilities and companies that treat, store, or dispose of wastes do not have to report under TRI, as of the writing of this book.) An environmental annual report allows a company to provide additional explanations to its TRI data, along with information on corporate goals, upgrades to environmental controls, cleanup of contaminated property, and other information that gives a more complete picture of the company's efforts to protect human health and the environment.

Most environmental annual reports produced today are expensive, professionally prepared pieces, and although they may provide detailed information, including summaries on specific company facilities, these reports still focus primarily on major, corporate-wide issues. Thus, they are useful primarily as supplemental materials in the community relations effort.

Some companies also prepare environmental annual reports for specific facilities. These may be trimmed-down models of the corporate report, or they may be documents that facility personnel themselves prepared in-house to explain TRI numbers and describe facility environmental policies and initiatives. Unlike the corporate environmental annual reports, facility-specific environmental annual reports can be important documents for sharing information with interested residents and public officials in a community relations program.

INFORMATION REPOSITORIES

Information repositories are collections of documents about a site, facility, or project that are available to the public for review upon request. Programs such as Superfund require that information about contaminated sites be made available to the public through information repositories. Permit applications under RCRA must also be publicly available—usually in the community in which the facility applying for the permit is located.

The repository should be publicly accessible, preferably during hours that would make the information readily accessible to residents with a variety of work schedules. Public libraries and municipal and county buildings are typical locations for information repositories. Care should be taken to ensure that repositories are accessible to the people who are most affected—usually, the neighbors. Although this may sound self-evident, it may take some thought to find the best location for a repository for maximum convenience. For example, residents of inner-city neighborhoods may not wish to travel several miles by public transportation to a main library in another part of the city. In order for these people to have access to information, a local location, such as a school library or other city building, would have to be used.

The repository concept can also be used by community relations specialists in conjunction with voluntary community relations programs for facilities, projects, or sites undergoing voluntary cleanup to make information accessible to community residents and other stakeholders. For example, a facility may provide the local library or municipal or county offices with general information on facility operations, including fact sheets, brochures, environmental annual reports, and newsletters. Copies of permits, TRI data, work plans for site investigations or cleanups, and analytical data are also important documents to include in a repository.

Repositories that provide comprehensive information convey a strong message of openness—the information is available for anyone to look at anonymously. A common impediment to a dialogue is the unwillingness of stakeholders to personally approach managers to ask for information. This obstacle can be removed by making information available in a neutral location.

Repositories should be maintained on a regular basis both to ensure that any new materials are included and to replace any materials that become worn or are removed or mislaid. As always, repository materials should include the names and phone numbers of persons who can be contacted with comments or to obtain additional information.

When discussing voluntary repositories for facilities, the question of proprietary information and protection of facility operations often arises. Information included in permit applications which is marked "proprietary" is not placed in repositories—it need not be placed in voluntary repositories, either. The security of facilities—especially those that handle or manufacture hazardous materials—is an important issue. Facility managers should use their judgment about potentially "sensitive" information, and place in the repository only those materials that they are comfortable with the public seeing. Most of the types of information that interested members of the community would want to see would not fall into this "sensitive" category. If a member of the public wishes to obtain more detailed or potentially sensitive information, he or she could make a request to the facility manager, who could determine whether the request should be met.

WORKING WITH THE MEDIA

Community relations specialists will need to work with the media in conjunction with most community relations programs for at least three reasons:

- The media constitute important channels of communication for reaching stakeholders and the general public.
- The media provide valuable "third-party" viewpoints and reporting that can help broaden the appeal of the communication effort to residents who may be less willing to look at materials disseminated by the organization responsible for dealing with the issue in question.
- Environmental issues are news. Failure to work in a positive fashion with the media won't keep stories out of print or off the air.

The media represent important channels of communication for community relations programs. As discussed in chapter 3, proper handling of the media is extremely important during crises. The media also figure prominently in reporting news about other non-crisis situations, including coverage of public meetings, TRI data, and activities surrounding the investigation and cleanup of contaminated sites.

Making an effort to work with the media will at best result in stories that further the goals of the community relations program. At worst, such efforts should at least minimize the amount or severity of negative reporting about the issue.

Get to know key reporters at the outset of a project or the implementation of a community relations program. This can be accomplished through arranging to drop by the newspaper, radio station, or television station to introduce yourself to the reporter, editor, or news director responsible for handling stories about specific issues (e.g., environmental, neighborhood, business). Explain the goals of the community relations program, and why it will be necessary to keep community residents engaged in a dialogue and informed of developments. The community relations specialist and the project or facility manager should be included in the meeting. Background materials, such as fact sheets, news releases, brochures, and the like, can be left so that the reporter can start a file on the subject. Reporters can be invited to visit and tour facilities. They can also be invited to visit a site undergoing investigation or cleanup if there is an appropriate meeting place away from the contaminated area.

Meeting with media representatives in advance creates an opportunity to establish a relationship and to provide background information on both the issue in question and the community relations program and its goals. Such meetings benefit media representatives because they provide them with the names of contact persons from whom they can obtain additional information as the issue unfolds, as well as a more well-rounded view of the issue. This invariably raises the quality of reporting on the story as it progresses and increases the likelihood that the information that is reported will support the goals of the community relations program.

For example, if the community relations program is seeking to obtain community input into and ultimate acceptance of cleanup options for a contaminated site, working with the media to ensure that they publish or broadcast good descriptions of cleanup options and their advantages and drawbacks, *as well as* the names and phone numbers of contacts that residents can call to make comments or obtain more information, will do more for the goals of the program than general stories that simply report that options for cleanup are being explored.

When working with the media, one must be sensitive to reporters' deadlines. Be sure to ask reporters what their deadlines are so you can give them the most current information in time for inclusion into their stories. Additionally, persons who are likely to be working with the media—or who may be called upon to act as spokespersons during crisis situations—should undergo media training. Media training includes exercises and role playing that teaches spokespersons how to deal with reporters and answer their questions clearly and credibly while making the points that are important to further the goals of the community relations effort.

Newspapers

Newspapers offer several possibilities for disseminating information. News and feature stories about the site, facility, issue, and so on are the most obvious. Paid advertisements can also be effective. Editorials and letters to the editor are additional possibilities.

News and Feature Stories   As discussed at the beginning of this section, community relations specialists may wish to initiate stories to provide information to the community. As we have discussed, the best way to do this is to establish a relationship with a reporter or editor and then to provide him or her with press releases and other information as the situation progresses. Reporters and editors are also likely to initiate stories themselves about an issue. The community relations specialist will want to work with them to ensure that the story accurately reflects the situation. Again, we stress, if you *don't* provide information, reporters will find someone who will—and the information they provide may be detrimental to the community relations effort.

News and feature stories are typically prepared by newspaper reporters. Thus, community relations specialists have limited control over what is written or how it is written. As a rule, reporters do not show drafts of their stories to the persons they interview. As a result, community relations specialists who are working with reporters need to be sure that they are providing information that is readily understandable to reporters and that they are available to provide clarification if needed. Providing reporters with written materials, such as fact sheets and news releases, can increase the likelihood that the story will be accurate. Additionally, as discussed in chapter 3, reporters typically develop their stories by talking to a number of different sources. Thus, if an activist, neighbor, or public official makes critical statements, these statements are likely to appear in the story. Community relations specialists can frequently have the last word, however, by anticipating what the other party will say, and offering a "proactive rebuttal" to the reporter.

In one case, a community relations specialist for a facility could guess from past experience that an outspoken activist in the community would state that the facility had received "lots of fines." The facility *had* received fines; however, most had occurred almost six years before, all had been settled, and recent inspections of the facility had been "clean" as a result of a major program to bring all areas of the facility into compliance. Although the story did include the anticipated quote from the activist, one of the angles of the story was, in fact, the improvements in the compliance record of the facility— a positive story.

The community relations specialist had anticipated the activist's comments and had provided the updated information. He caught the attention of the reporter, when providing this information, by saying: "You may have heard that we had some problems here with fines. That's true, we did. Now let me tell you how we did on our last three inspections. . . ." He also suggested that the reporter talk to an inspector in the fire department who was well acquainted with the facility and had worked with its management to ensure compliance with the municipality's fire codes, and provided the name and phone number of the state department of environmental quality inspector, as well.

It is important to take control of the situation when reporters call for information. Ask them what the story is about so you can provide complete responses and detect whether the story is headed in a damaging direction. Since environmental issues can easily be misinterpreted, providing answers to a few isolated questions about the situation at a site or the environmental performance or practices at a facility can lead to erroneous or misleading information in the newspaper.

In one case, a reporter asked for permission to come onto a facility's property to photograph an eagle's nest in a wooded area behind the plant adjacent to a nearby river. Facility management was pleased to have the publicity—after all, it demonstrated how good the wastewater treatment system was at protecting wildlife in the area. Since it had been upgraded, the facility's wastewater discharge met drinking water standards—cleaner than the river it was discharging to. In fact, the facility had upgraded all of its pollution control systems.

Unfortunately, the one thing that facility management *hadn't* done was communicate effectively with the community—or with reporters—about how much its environmental performance had improved.

When the picture appeared in the paper—a picture of the eagle's nest with the facility's stack in the background—the caption stated that the eagles were nesting despite the pollution being caused by the facility.

Advertisements   Since community relations specialists have limited control over news and feature stories in any of the media, paid advertisements are increasingly being used to provide information to the public. Advertisements print the exact words that community relations specialists wish to convey and can be scheduled to appear on the date or dates desired. Relative placement of advertisements in the newspaper (or, in the case of broadcast advertisements, time of day or program slot) can also be stipulated. In the case of print ads, graphics, such as maps showing the location of a site undergoing an investigation or a facility undergoing permitting, can also be included.

Advertisements may be used to announce the dates, times, and locations of public hearings or meetings or the availability of information on remedial alternatives. They may also be used to provide information or to issue statements regarding an issue, incident, or situation.

For example, as part of the communication in the aftermath of a chemical release at a facility that resulted in the evacuation of a number of area residents, the facility ran a full-page ad in the local newspaper. The advertisement included an apology to the community for the fear and inconvenience caused by the release. It provided information on immediate steps the facility had taken and was taking to investigate the incident and identify the cause of the release. The advertisement also stated that the results of the investigation would be made public, and provided a contact name and telephone number that people who had questions or concerns could call for additional information.

Obviously, running an advertisement is only one part of the effort needed to communicate with the public about an issue such as a chemical release. Use of an advertisement ensured timely distribution of the information, however, and provided a very public forum for the facility's apology to the community.

Advertisements may have limited effectiveness for changing the attitudes or opinions of people regarding a controversial subject. Their very nature, as pieces paid for by the party on whose behalf they are communicating, means that the public tends to view advertisements as biased and self-serving. This doesn't hamper their use for announcing events, such as public meetings, or for providing emergency information, such as phone numbers and contact names, as in the case of the release. It does mean, however, that advertisements intended to rebut misinformation or to persuade the public regarding a specific issue are less likely to have the desired effect—and, in some cases, may create a public backlash against what is viewed as an attempt to "buy" public opinion.

Some organizations use advertising independent of specific issues to communicate about their philosophy or support for the community. For example, organizations may buy ad space in school yearbooks or to support charitable causes. Such advertisements aren't going to override serious community concerns; however, they may be a legitimate component of the community relations program's overall effort to build a positive image—and an image of involvement and accessibility—in the community.

Editorials and Letters to the Editor   Editorials are typically written about issues that have already been in the news. Thus, a positive editorial will most likely come as a result of earlier work with the media. As in the case of issues statements, editorials are interpretive, explanatory pieces that look at the total effects of policies or actions. Editorials can provide important third-party endorsements in cases involving complex, high-profile, or serious issues or situations. Depending on the stature of the newspaper—and individual stakeholders' opinions of it—a positive editorial can provide both validation of the effectiveness of community relations activities and, often, closure to certain issues.

Of course, situations or issues that are mishandled in terms of community relations, or that have a very serious impact (or perception of an impact) on a community or ecosystem, can result in negative editorials. These can summarize, amplify, and legitimize community criticism.

Negative editorials rarely appear out of the blue—although editorial opinions are by no means always correct. If persons in charge or involved with an issue are receiving a lot of bad press or community criticism, they need to look at their activities—or lack of activity—and at their community relations program to determine whether changes are needed to address community concerns. A meeting with reporters or with the editorial board should be considered to discuss with them the measures that are being taken both to address community concerns and to fix the problem. If there are no actual technical problems or detectable adverse impacts to human health or the environment, this also needs to be made clear, along with the plans to work with the community to get at the root of residents' concerns.

Negative editorials are less likely to appear when positive dialogues are under way—even if significant difficulties in correcting a problem (e.g., cleaning up a spill, remediating a site) are hampering progress. As in the case of crisis communication, editorials are as likely to focus on the way an organization is dealing with an issue or handling community concerns as they are to focus on the issue itself.

Community relations specialists should consider responding to both positive and negative news or editorial coverage with one or more letters to the editor. Letters to the editor should be short and to the point (preferably no more than 50 or 60 words). Letters mailed within a day or two of publication of the story they are commenting on are more likely to be published than those that arrive later—after the news is "old."

Although not all letters to the editor are printed, their preparation and submittal is still important, since they convey an interest in maintaining a dialogue with the media about a site, facility, or project. Thus, even those that are not printed may affect future coverage of an issue by providing additional information (which may be used for additional story ideas) and by reinforcing the writer's willingness to serve as a source of information.

In the case of positive coverage, the letters should thank the newspaper for coverage of an important community issue and provide some additional information that wasn't included in the story (invariably, there is always something that was omitted because of space considerations).

- A facility manager may write that the facility's successes in pollution prevention (the subject of the story) is part of a total program that seeks to enhance environmental quality in every aspect of its operation to benefit the community, as well as overall corporate performance.
- A project manager for a site undergoing an investigation may write thanking the newspaper for covering this important issue (the investigation), adding that it is the project team's intent to ensure that the community be kept abreast of developments at the site, and providing the name and phone number of the contact person for the site if it was not included in the story.

As the reader will note from these examples, the community relations specialist often isn't the one who signs a letter to the editor—even if he or she writes it or provides an outline of what should be said to the person who ultimately signs. Community relations specialists may also encourage or help key stakeholders to write letters to the editor.

Letters to the editor are also important responses to stories that report erroneous or unduly negative information. Such responses should be measured and objective, with an emphasis on clarifying misunderstanding rather than on condemning reporters' skills or the editorial board's perceptions.

Broadcast Media

Although newspapers are the most common mass medium used in community relations activities, radio, and even television, are important alternatives. The community assessment should determine the types of media from which stakeholder groups receive information. The major daily newspaper may not be the best way to reach certain stakeholders. A weekly "neighborhood" paper may be a better bet to reach some stakeholders. Racial or ethnic minorities, or persons whose primary language is not English, may have their own newspapers or radio stations. Radio is a highly segmented medium, and specific stations often have narrowly targeted audiences. Broadcast media are frequently willing to make public service announcements about upcoming public meetings or the availability of information about a pending permit or remedial alternatives at a site in the community if these announcements are presented to the station manager as being of direct interest to the station's audience. Paid advertisements for events, such as open houses, can also be used effectively on local radio or television stations.

Both television stations and radio stations offer local broadcasting possibilities beyond public service announcements, advertisements, and news. Many stations also air talk or discussion shows that feature local issues. Since environmental issues can make for lively discussion and are of interest to large segments of the viewing or listening public, community relations specialists can often get their issue on the schedule.

Particularly in the case of television, there can be substantial risk to "going on the air" to discuss an issue, and the services of a good media relations specialist should be retained to help ensure that the experience will be a positive one. Significant attention should be paid to the format of the show and who the other guests might be. For example, appearing on a local science show with a set format and a knowledgeable host to explain about groundwater remediation will be less risky (but probably receive fewer viewers) than appearing on a talk show to debate chemical use with local activists. Opportunities to appear on broadcast media must be carefully assessed—with assistance from a media relations specialist—on a case-by-case basis.

News Releases

News or press releases are the most common method for providing information to the media. They should be written to convey information that is newsworthy, such as receipt of analytical results during an investigation, submittal of a permit application, plans for a plant expansion, or the beginning of a new phase of a development or anticipated development. News releases can encourage editors and reporters to provide coverage of a story or provide additional information on a story that is already being covered. In some situations, a news release may be picked up, either in its entirety or in portions, if it is objective and follows the conventions of news writing.

News releases should be written in the same manner as news stories—with information presented in descending order of importance, since news stories are typically cut from the end forward to fit available space. The lead sentence is the most important element of the press release. It should clearly convey the theme of the story, preferably in 20 words or less. As much as possible, the lead should convey the impact (or lack of impact) on people or the environment, since this is the "hook" that typically makes a story newsworthy. For example:

> ABC Company announced today that it will pay for city water hookups for three households in Westbrook. The action was announced after results of an investigation indicated that chemical compounds from a leaking underground storage tank at ABC's Bay Road facility were present near the facility's west property line. . . .

News releases should be written objectively and avoid editorializing.

In addition to following news story style in terms of writing, news releases should look as professional as possible. For ease of editing, news releases should be double-spaced. News releases should always include a name and phone number of a contact person at the top right-hand corner of the first page. They should be marked "news release" and include a short, tentative title for the story.

Most news releases will be relatively short—from one to three pages. Some releases, such as those about complex legal or regulatory settlements or the results of an investigation into an accident, may be considerably longer. Such lengthy news releases are most often prepared to act as backup materials for distribution at a news conference or briefing to ensure that reporters get a good handle on complex issues. Additional backup materials may also be distributed along with news releases at such briefing sessions.

Nowadays, news releases are often sent via facsimile machine to media newsrooms. Some reporters also have e-mail addresses to which correspondence can be sent. Since newsrooms receive large amounts of information, a call to the assignment editor or to the reporter who has been following a story, or who regularly covers a beat that the story would fall under (e.g., business, environmental, neighborhood news), to alert him or her to the availability or pending arrival of the news release is a good idea. If a news release is being issued on a story that is of state, regional, or national interest, then it is typically released via newswire, with calls placed to alert key media representatives to the availability of the information.

News Conferences and Briefings

News conferences are generally held to make important announcements about major issues. The term "news briefing" typically refers to updates on ongoing situations. Both of these events are typically interactive sessions, in which a statement is read at the beginning of the meeting and is followed by a question-and-answer session. One of the purposes of a news conference or briefing is to allow the release of important information to all media representatives at the same time. As discussed in chapter 3, in cases of crisis communication or when an issue is of pressing and immediate importance to the community, it is important not to "play favorites" by preferential release of information to some media representatives but not to others. (Note: One should still build relationships with individual reporters, as we've just discussed. During a crisis, however, or in the case of announcements important enough to warrant a news conference or briefing, don't play favorites.) Attention should be paid to reporters' deadlines when picking the time for a news conference. If different media have different deadlines and the story is one that is unfolding over a period of time, you might consider alternating briefing times to accommodate the deadlines of different media.

Background information on the facility, site, or project, as well as current fact sheets and news releases, should be provided to reporters to help them prepare accurate stories. Announcements that a news briefing is going to be held should be conveyed to all media representatives who are in the area or who are likely to be interested in the story. News conferences or briefings should be held in rooms that can comfortably accommodate the number of media representatives (and technicians, in the case of broadcast media) who are likely to attend.

Since news conferences will involve several—and, in some high-profile cases, many—reporters, they must be carefully managed. As much as time will allow, formal questions and answers should be prepared and rehearsed to ensure consistent, easy-to-understand, accurate answers. Most questions from the media can be easily anticipated by a careful review of the issues that have come up thus far, knowledge of the subject and its history, coverage of similar news events, and community sensitivities identified during the community assessment. Experienced community relations specialists can often script out the entire news conference with amazing accuracy beforehand.

Since reporters invariably look for multiple sources, it is desirable to have the "experts" who are working on the issue available during the news conference. In such cases, assignments as to who answers what types of questions should be clearly drawn and rigorously adhered to. For example, the hydrogeologist who just answered a question on groundwater flow should not attempt to answer a question on risk if a doctor or risk analyst has been brought in to answer such questions. Media representatives often don't know who should be answering technical questions and many address their questions to the wrong person, especially if speakers are taking turns on the podium. One way to ensure that the right person responds is to have the person running the news conference direct the questions to the appropriate party.

News conferences can sometimes be contentious, especially if the issue is serious. Personnel who are speaking or providing information should remain calm and measured in their responses. Displays of anger or extreme nervousness will show up on TV news. If possible, persons who are likely to find themselves in the position of dealing with the media should undergoing media training—with regular refreshers to help their performance.

Editorial Board Briefings

When an issue is high-profile or involves a long-term situation that is likely to be of interest or concern over a significant period of time, community relations specialists may wish to arrange for an editorial board briefing to provide the newspaper's editors with additional understanding of the project and the community relations approach. The briefing would provide information on the issue in question and explain the intent of the community relations program and the importance of involving stakeholders.

Although it is also desirable to share this information with reporters so that they will write better-informed news stories (and we recommend doing so informally), editors are the ones who assign the stories and determine what is going to be printed. By providing information to the editors, stories that help meet the needs of the community relations program are more likely to be assigned in the first place. Additionally, it should be noted that reporters don't typically write the headlines—editors do. It is not uncommon for a story that is positive, or that conveys important messages that are in keeping with the goals of the community relations program, to be undermined by a headline that reflects common, though incorrect, opinions about the issue in question. For example, a positive story about a facility's pollution prevention initiatives may end up sporting a headline such as "Local plant emits toxins."

Editorial board briefings aren't necessarily intended to produce an immediate positive editorial about the issue. Rather, they are intended to provide background information to the editorial board as to why the newspaper should provide coverage about an issue or to explain the type of coverage that the newspaper should provide to meet the community relations goals of working with the community. Editorial board briefings may also be requested to address misunderstandings by the media about an issue.

For example, if newspaper stories have been critical about the length of time that an investigation or cleanup is taking, a briefing to describe the procedures that are followed to ensure a thorough investigation can help editors understand the investigation and remediation process. This should limit the amount of negative reporting about at least some aspects of the investigation or cleanup, and increase the likelihood that editors will choose to aid the community relations effort by printing stories that provide the type of information that the public needs in order to participate in a dialogue regarding activities at the site.

EVENTS

Another common category of community relations activities would best be described as "events." The events we will discuss here include public hearings and meetings, small group meetings, workshops and availability sessions, open houses, tours, and exhibits that provide immediate, face-to-face opportunities for people to learn about the issue in question. Events are important because they provide the opportunity for two-way communication with stakeholders. Since they involve physically going somewhere, they also have the capacity to engage the attention of stakeholders much more effectively than written materials or information reported in the media.

Events such as public meetings and open houses require significant planning and effort to be effective. When done properly, however, they can be powerful tools for promoting a positive dialogue with the community.

Public Hearings and Meetings

The term "public hearing" is typically used to refer to events that are arranged and conducted by a governmental agency to meet statutory requirements, although some agencies may use the term "public meeting" to refer to these events. Companies may also be required to hold public meetings under certain circumstances or to meet regulatory requirements. Public meetings or hearings may be held during the RCRA permitting process, to present remedial alternatives under CERCLA, or in conjunction with zoning decisions or projects involving transportation or the use of public lands. Public meetings or hearings convened by government agencies typically involve presentations by agency personnel, and often include opportunities for the public to make comments or ask questions. Comments or questions from the public may or may not be answered during the meeting.

Company representatives may also be asked—or allowed—to make presentations at such government-convened meetings. In most cases, it is desirable for companies to make presentations and to answer those questions that can be answered without additional research. To refuse to speak—or to answer questions—at a public meeting flies in the face of the concept of an open dialogue. In the case of permitting, zoning, or remediation plans, company representatives may also want to encourage other stakeholders or officials to attend the meeting and to speak out on behalf of the company. Such third-party endorsements can provide a powerful boost to public confidence regarding the management of a facility, site, or project.

The "down" side of public hearings or meetings is that they can turn into angry, unruly sessions that can occasionally erupt into violent protest or confrontation. Savvy community relations specialists and agency personnel realize that large public meetings are *not* appropriate forums for hashing out serious concerns or disagreements. Thus, steps should be taken to answer questions and resolve community concerns *before* the formal public meeting takes place. Small group meetings, availability sessions, and workshops, are

---

Using Total Quality Management Principles in Public Meetings

John Perrecone, community involvement coordinator in the Office of Public Affairs, U.S. EPA, Region V, suggests that total quality management (TQM) principles can and should be used in communication with the public. Public meetings, though required for many kinds of community relations efforts, can often present a rather ineffective forum for real communication. Those with an agenda can sometimes dominate the proceedings, while others may simply feel uncomfortable expressing concerns in such an authority-laden format.

TQM techniques can be used to enhance the communication process by encouraging participation and establishing a positive dialogue. The meeting may open, as usual, with a discussion of the meeting's purpose and agenda. Then, rather than continuing to the expected presentation, the moderator can open up the floor to discussion about the agenda and ask the audience for its input. Depending on the responses (which should ideally be written down where all can see and confirm what has been said), the meeting can continue with the presentation as planned or be amended to incorporate the issues raised during the open forum. Thus, the audience members feel that they have had a participatory role in the proceedings, and that their questions have been heard and taken seriously.

tools that can be used, along with neighbor visits and briefings with local officials and other stakeholders, to answer community questions and work out concerns before a formal public meeting is held.

Small Group Meetings

As the name implies, small group meetings involve small groups—typically 10 to 20 people, at most. Often, a series of meetings is held in order to reach as many stakeholders as possible. These informal meetings involve in-depth discussion of concerns and issues, and they allow stakeholders to ask questions and receive answers more easily than in a larger, public forum. They are also much less likely to devolve into mob scenes if emotions run high. Small group meetings may be held by government agencies or by private sector organizations.

Project or facility managers and community relations specialists should be present at these meetings. Other technical personnel or public officials, such as emergency response personnel, can also be asked to attend the meetings if the types of questions or concerns that participants are likely to have would benefit from their input.

Small group meetings may be held in someone's home; in a public building, such as a library; at a facility (if community members are willing to go there); or anywhere else conducive to frank discussion that would suit the participants. One potential criticism of small group meetings is that the organizations holding them may be accused of avoiding confrontation with the community as a whole, or of providing different information to different groups. As the EPA's manual *Community Relations in Superfund: A Handbook* (EPA/540/R-92/009; NTIS No. PB92-963341, January 1992) warns, the organizations holding small group meetings (whether public or private sector) may be perceived as attempting to "divide and conquer." One technique suggested in the EPA manual is to keep a written record of the discussions, which can be made available upon request. Another technique used with success by community relations specialists is to have a well-respected, neutral party in attendance or acting as a facilitator at all of the meetings who can vouch for the fact that the same information was provided to everyone.

Determining how small group meetings should be structured takes careful consideration. For example, should the groups contain a cross section of stakeholders? Persons from specific neighborhoods or stakeholder groups only? Or whoever wants to come after publicizing that a series of informal meetings will be held? Should they be held through the sponsorship of local civic or community organizations? Should the media be invited to attend? The decision will depend on the issue itself and the way community members are responding to it. Another consideration, if public officials are going to attend, is whether the state or municipality has "sunshine laws" that prohibit officials from participating in meetings that are closed to the general public. In such cases, public officials could participate only in meetings that were open to all community members.

As always, thorough preparation by persons who will be answering questions or making presentations is of the utmost importance. Model questions and answers should be developed to ensure consistent, easy-to-understand answers. Fact sheets and other information should be available for meeting attendees to take home with them. If such information hasn't been developed yet, then business cards or sheets of paper containing the names, addresses, and phone numbers of contacts should be provided. Additionally, although minutes or transcripts often won't be prepared for such informal meetings, it is important that a record be kept of comments and questions so that, if necessary, additional materials can be developed to address them. If a question cannot be answered during the meeting, it should be researched and an answer should be provided after the meeting.

Persons who attend formal meetings and hearings are typically required to provide their names and addresses, both so that any questions that weren't answered during the meeting or hearing can be answered in writing afterward and so that their names can be added to the mailing list for subsequent communication on the issue. If possible, names and addresses should also be collected from persons who attend informal small group meetings, for the same reasons.

Voluntary Public Meetings

Organizations may wish to hold voluntary "large group" public meetings to provide a forum for the exchange of information. If knowledge of the community and of the issue indicates that such a meeting will not turn into a negative, mob-driven experience, this is perfectly fine. Some communities have a long tradition of public meetings and would view small group meetings as suspect or unnatural. The rules for these types of voluntary large group meetings are similar to those for small group meetings in terms of preparation, availability of contact information, and fact sheets. Extra effort will have to be put into graphic materials, such as poster boards of site plans, to ensure that attendees can see them from far away.

Voluntary large group meetings can also be appropriate to reinforce positive relations with the community. For example, some manufacturing facilities and public sector entities, such as parks or forest preserves, hold annual "reports to the community" to provide information on events or performance during the past year and plans for the future. Such events can be very well attended in some communities. These types of "maintenance" activities are part of an ongoing community relations program designed to keep channels of communication open.

Availability Sessions and Workshops

Availability sessions and workshops are used when complex information must be conveyed about an issue that is the subject of serious concerns. Both workshops and availability sessions are excellent, if labor-intensive, methods for providing stakeholders with the types of information they need to be able to participate fully in the dialogue. Availability sessions involve allowing stakeholders to talk with technical experts, such as doctors, academics, geologists, remediation specialists, and emergency responders. Although they often include brief presentations, the goal of the availability session is to allow stakeholders direct access to knowledgeable people so that they can have their questions answered. Availability sessions are typically free-form events held in public facilities to encourage stakeholders to explore issues that are of concern to them.

Availability sessions are useful when stakeholders have enough knowledge, interest, and concern to articulate their questions. Sessions that provide direct access to third-party experts can also be especially useful in cases in which stakeholders distrust the company or governmental agency that is responsible for the issue in question.

The key consideration is whether stakeholders are willing to attend an availability session. If the community assessment indicates that important stakeholder groups would not be likely to attend such sessions—for example, if some stakeholders have limited formal education and indicate that they would feel uncomfortable questioning technical experts—then some other activity, such as an informal small group meeting, where information is presented to them, is likely to be a better choice to reach these groups. On the other hand, highly educated stakeholders or stakeholder groups in communities that are very inclusive and interactive are likely to welcome the opportunity to have direct access to technical experts without having to sit through long presentations.

Since availability sessions involving technical experts (especially third-party technical experts) take substantial amounts of effort to arrange, they should be well publicized to ensure good attendance. Invitations to key stakeholders should include descriptions of the experts who will be available for discussion.

Workshops can be compared to mini-courses or seminars on specific technical or regulatory topics. Workshops typically last for several hours and are used to provide an understanding of the issue in question. For example, a workshop may be devoted to exploring the RCRA permitting process so that key opinion leaders and other interested parties can understand the application process, the aspects of an operation that are covered by the permit, how the permit is issued, and how the permit conditions are enforced through inspections. Another workshop might be aimed at educating stakeholders about groundwater—what it is, how monitoring wells are installed and sampled, or how contaminants can be removed from it by using a pump-and-treat system.

Workshops are typically targeted at primary stakeholders or opinion leaders who are formally invited. Workshops can be open to the general public, as well, if interest in the community is sufficient. Such "open" workshops should be well publicized. The number of people who can be accommodated in a workshop will depend on available space, community interest, and the type of information that will be presented. For example, if the workshop will involve looking at or operating instrumentation and providing some "hands-on" experience, then the number of participants will necessarily have to be limited.

Workshops take a significant amount of effort to arrange, since printed materials, presentations, and graphics, such as posters, must be prepared. As in the case of availability sessions, some workshops benefit from having third parties teach them. The efficacy of workshops can be measured with pre- and post-tests—a good mechanism for helping the attendees realize what they didn't know and how much they have learned.

Tours and Open Houses

Tours and open houses can offer opportunities for firsthand observation of a facility, site, or project. In some cases, an open house or tour may not be conducted at the actual facility, site, or project in question. For example, a tour conducted in conjunction with the discussion of remedial alternatives for a contaminated site may be held at another, similar site that is undergoing—or has undergone—one of the remedial alternatives under consideration.

Additionally, open houses held in conjunction with efforts to find a location for a waste disposal facility, for instance, may take place at a hotel or public building, since the facility hasn't been built yet. The open house would allow members of the public to view exhibits related to the proposed facility and speak to persons who are either involved in the process of finding a location for the facility or will be involved in building or managing the facility. Thus, even in cases in which the actual facility or site is not available for open houses or tours, holding these activities at other locations can still be an important form of outreach to the public.

Tours and Open Houses at Locations Other than the Facility or Site As discussed in the previous paragraph, it may be necessary to hold tours or open houses at locations other than the actual facility or site. If the issue in question involves siting and building a facility, then touring a similar type of facility can help stakeholders envision and understand what is being planned for their community. Especially in the case of waste treatment, storage, and disposal facilities (TSDFs), it may be difficult for stakeholders to envision what the proposed facility will look like, or to understand how the safeguards that are built into it really operate.

Tours should include presentations that will provide enough information so that participants gain a good understanding of what is being done, why it is being done, and how operations are designed to protect human health and the environment. Fact sheets and other background materials that highlight and supplement the information provided during the tour should be distributed to participants. Tours should include plenty of time for questions and answers.

Tours should be scheduled to ensure that key stakeholders will be able to attend. Several tours may be necessary if a large number of people want to participate, since the number of participants on each tour should be limited to those who can be reasonably accommodated by the physical facilities and hear what presenters are saying as the tour progresses. Unless the facility or site that is going to be toured is located nearby, participants should be transported to the facility or site in a hired bus or van, both to ensure that participants arrive together so that the tour can commence on time and to remove the burden on participants of having to drive to the tour site themselves.

Tours may be by invitation only or they may be publicized, via newspaper, radio, announcements at civic organization meetings, and so on, as open to anyone who is interested. If the latter course of action is chosen, it is wise to request that persons who are interested in attending call ahead and make a reservation so that organizers will know how many people to expect and can schedule additional tours if many people wish to attend. Tour partici- pants should be given forms or postcards that they can fill out and either hand in or mail in with any additional questions or suggestions they might have that weren't addressed during the tour.

Off-site open houses are similar to availability sessions with exhibits. The goal of these types of events is to bring interested members of the public together to talk with persons involved with development or operation of the facility, site, or project and view posters and exhibits on the subject without the structure of a public hearing or public meeting. Written information, such as fact sheets, should be available, as well as sign-up sheets for visitors who wish to request additional information or who wish to be added to the mailing list about the facility, site, or project.

Open houses are an excellent tool for initiating dialogue with members of the public; however, attendance is often less than organizers would like. Potential dates should be carefully checked to make sure they do not conflict with local holidays or other important events. For best attendance, open houses should be well publicized through multiple means well ahead of the date selected. This will probably mean a mix of paid advertisements in the newspaper and/or on the radio, personal invitations to public officials and other key stakeholders, newspaper stories, posters in public buildings, and announcements at community organizations. Although the goal is for commu- nity members to attend the open house and contribute to the dialogue regard- ing the facility, site, or project, the publicity surrounding the event can also help educate community residents about the issue and the availability of information about it.

Tours and Open Houses at Existing Facilities Facility tours are a very important part of an "open door" policy and should be well publicized. Even in cases in which many community residents never bother to tour the facility, knowing that the option is there communicates an important message to residents. Every time a representative of facility management meets with members of the community or speaks at a school or community group, he or she should extend the offer to tour the facility.

Both public and private sector facilities use tours and open houses to reach out to community members. Holding tours or open houses at existing facilities is similar to the off-site tours and open houses we just discussed. The following paragraphs will discuss the differences.

Many public and private sector facilities that manage wastes, provide public utility services, or engage in large-scale manufacturing are toured on a regular basis. In the case of facilities that manage wastes, tours by customers or prospective customers are a standard part of customers' "due diligence" to ensure that their wastes are being managed properly. Tours are so common at some facilities that they are routinely scheduled several times a week.

As already discussed, tours should include presentations on important aspects of the facility's operations. Facilities that are subjects of frequent tours often develop sophisticated programs. Handout materials providing additional information or overviews of what is in the tour should be provided. As always, these materials should include the name and phone number of a contact person. Because of their educational nature, facility tours can make excellent field trips for school children and for community organizations. Inviting these groups to tour is an excellent way of building familiarity with community residents.

On-premises open houses are similar to off-site open houses in the amount of work that must be done to prepare for and publicize them. Special attention should be given to inviting both commercial and residential neighbors. Employees and their families are also important attendees, especially if they live in the community.

Open houses at existing facilities are typically festive events, including games, food, door prizes, and special demonstrations—such as the use of emergency response equipment. They also include more serious, education-oriented content, such as tours of the facility.

Exhibits

Exhibits illustrating the investigation and cleanup process (or certain aspects of it), the permitting process, wetland mitigation, or any number of other subjects can help educate the public about environmental issues. Since exhibits include both text and graphics, they are especially useful for showing multistep processes and explaining complex natural cycles. Some more sophisticated exhibits make use of film loops, slides, audio tapes, and interactive computer programs.

Exhibits can be placed in public buildings, such as libraries or municipal centers, for continuous, unattended display over a set period of days or weeks. When exhibits are unattended, they constitute one-way communication; however, the presence of graphics and the overall layout can make this one-way communication tool more effective than one-way communication on a fact sheet, because it is likely to engage the viewer more fully. If written materials, such as fact sheets, are provided at the exhibit for interested persons to take with them, supplies should be replenished on a regular basis.

Exhibits can also be set up in conjunction with public hearings and meetings or at community events, such as chamber of commerce fairs or county fairs. When exhibits are set up for display at events, they should be attended by persons who can answer questions and provide additional information. Fact sheets or other supplementary materials that contain a contact name and phone number should be made available at the exhibit. As in the case of off-site open houses, a sign-up sheet should be available for persons who would like to receive more information. Feedback forms soliciting questions or concerns should also be available at the exhibit.

OTHER MAINTENANCE ACTIVITIES

As we said at the beginning of this chapter, community relations activities are typically best categorized not so much by the actual activity, but according to how they are used. For example, neighbor visits may be used to promote a dialogue on a specific topic or they may be used simply to "maintain" good relations, even if there is nothing much to talk about. Other activities, such as public hearings, are really only held to address defined, regulatory and technical issues.

In this section, we will look at some of the remaining types of activities that community relations specialists can use to help promote an ongoing dialogue with the community. We call this section "other maintenance activities" because these activities are often focused more on relationship and "image" building than on targeted communication about specific, serious issues. The one possible exception to this is the first activity—speakers bureau. Obviously, speeches can be targeted at providing specific information about a timely issue. We placed speakers bureau in this section because such activities are typically ongoing programs with primary goals of building familiarity with the organization as a whole—not simply the issue in question.

Maintenance activities are most applicable to facilities or projects of long-standing duration. In both of these cases, a long-term relationship and presence in the community is desirable. Maintenance activities offer opportunities for organization personnel to become known in the community and to extend their presence beyond specific issues.

Speakers Bureau

Speakers bureaus comprise individuals within organizations, public or private sector, who are available to speak on certain topics to schools, civic groups, business groups, and the like. Speakers bureaus actually do two things: They provide information to the community and build relationships and a positive image of the organizations they represent, *and* they help develop public speaking skills among organization employees. Both of these results are useful from the community relations standpoint.

Speakers bureaus typically have a coordinator who outlines, writes, or aids in the development of speeches on a variety of topics. Persons who volunteer to participate in the speakers bureau become familiar with one or more of these speeches. Speakers bureau members receive coaching on their speaking skills and practice their speeches so they can give them well. Summaries of topics that the speakers bureau can discuss are sent to officers of community groups and principals, department heads, or specific teachers at schools along with offers to speak.

An active speakers bureau program can often reach a wide variety of community residents. Speakers bureaus can build positive reputations and images of public-spiritedness for organizations that sponsor them. This can enhance other community relations efforts regarding environmental issues or make the establishment of a dialogue in the event of a newly emerging issue that much easier.

School Programs

Involvement in community schools may include direct communication, such as conducting tours of a facility or speaking on certain topics specific to a facility, site, or project. It may also include providing tutoring or coaching services or donating goods or equipment that have nothing to do with the issue that is the subject of the community relations effort, but that are important and necessary for the schools and the community.

This second type of school involvement—tutoring or donating goods or equipment divorced from a specific issue—is an exercise in good organizational citizenship that may not communicate about a specific issue, but can speak volumes about an organization's commitment to the community. This is not to say that an organization that is having—or is perceived as having—environmental problems can "buy" community acceptance by donating football uniforms or textbooks. Rather, this type of activity, especially over time, can help ease the dialogue on other issues because residents are likely to feel that the organization does indeed have some interest in the community. Benefits from working with the schools include positive publicity, contacts with school personnel and other formal and opinion leaders, and a positive image among parents whose children are the recipients of the extra help.

Community Organization Involvement

Both public and private sector organizations can build relationships with community residents through memberships and participation in community organizations. Regular attendance at meetings held by the local chamber of commerce, business organizations, environmental preservation organizations, and other such groups can help facility or project managers get to know some of the key opinion leaders in the community. Such involvement can also help facility or project managers gain an understanding of some of the current issues and concerns in the community.

Serving as an officer of a local organization does even more for visibility, and typically places the person who is serving "in the loop" to hear community news and make the acquaintance of decision makers or opinion leaders. Thus, serving or participating in appropriate organizations can enhance the ability to build the relationships that can help promote a strong dialogue—at least with some stakeholders.

Such outreach activities can including sponsoring community activities, such as a Little League team or a golf tournament to benefit a charitable organization, or participating in community activities, such as "cleanup days." These types of activities can promote a camaraderie that can build relationships that may be difficult to establish through other means. For example, corporate personnel may end up working with, and establishing good relations with, environmental or community activists whom they may not get to know other than within the context of a contentious environmental issue. Although, as we have explained, it is very possible, through the use of a positive dialogue, to build good relationships during the resolution of an environmental issue, beginning that dialogue from the common ground of a shared, positive experience can ease the process considerably.

COMMUNITY ADVISORY PANELS

Community advisory panels (CAPs) are groups comprising stakeholder representatives that are designed to help promote a dialogue between the stakeholders and the organization, such as a facility, whose issues they are exploring. (Similar types of groups—restoration advisory boards, for federal facilities, and community advisory groups proposed for use under CERCLA—are discussed in chapter 8.) CAPs provide a formal mechanism for the exchange of information and feedback. CAPs (which may go by other, similar designations, such as "communications committee," "community liaison committee," or "community advisory group") may be established in conjunction with facilities or with sites undergoing investigation or cleanup. For example, the Chemical Manufacturers Association encourages its member companies to form CAPs at their facilities as part of the Responsible Care® initiative. CAPs may also be formed by government or other civic or public entities to discuss area-wide issues, such as brownfields redevelopment or the use of natural resources.

When CAPs are properly designed and managed, they can provide powerful tools for enhancing an organization's dialogue with the community both by obtaining feedback from the CAP members that can be used to improve the quality of community relations activities and through the CAP members' own communication with other stakeholders. Since CAPs are necessarily small groups that meet specifically to discuss issues relating to a facility or site, they provide an excellent forum for the exploration of some fairly detailed issues. For example, CAP members may want information on the pollution control equipment at a facility. CAP meetings can function as mini workshops, with presentations by the environmental managers at the facility, facility tours, and question-and-answer sessions. CAP members have an opportunity to learn about the operation of a facility or about the procedures used to investigate a site in much greater detail than is typically possible through public meetings or fact sheets. CAP members are then in a position to communicate with others in the community about what they have learned, and to bring back additional feedback and questions. Despite the unique opportunities that CAPs offer for obtaining feedback from—and educating—key stakeholders, CAPs should supplement, rather than replace, other community relations activities.

Structure and Functioning of a CAP

In order to serve their intended function, CAPs need to include representatives from the stakeholder groups that have an interest in the facility, site, or project in question. When CAPs are criticized, it is typically because they are perceived as excluding important stakeholders or opinion leaders, either by design (e.g., the persons forming the CAP chose not to include representatives of groups that they believe may be disruptive, or they wrongly assumed that a group wouldn't be interested in participating) or because stakeholder groups were not identified. In either case, the efficacy of the CAP may be minimized. A community assessment will provide much of the information necessary to ensure that stakeholders are identified for the CAP.

In an interesting twist on CAP recruitment, some persons who are responsible for forming CAPs recruit the first few members and then have these members recruit the remaining members. Such an approach should be used carefully, since it could lead to the exclusion of stakeholder groups that the initial members don't know about, don't deal with, or don't like. A better alternative would be for the persons responsible for the formation of the CAP to map out the stakeholder groups based on the community assessment, recruit them, and then ask the new panel members if they have suggestions for additional members who might be included. This alternative approach will ensure that the CAP is well balanced, and it can bring in additional stakeholders or opinion leaders who the other CAP members feel should be involved.

One question that must be addressed in setting up a CAP is whether persons who are hostile to the facility or the organization responsible for the site or project that the CAP is formed to support should be included. Since the goal of a CAP is to promote open and frank dialogue with stakeholders who are representative of the community, it would defeat the purpose of the CAP to exclude critics; they should be included. Indeed, it is very difficult to have a thorough exploration of issues without some criticism and pointed questions.

If a potential CAP candidate is likely to be an extremely disruptive influence who would chase away other CAP members, however, persons responsible for forming the CAP would need to weigh whether this person holds a vital position as an opinion leader, and, thus, cannot easily be excluded, or if someone less contentious would be considered an adequate substitute. The decision to exclude a disruptive person who is a key opinion leader should be considered with the input of the CAP's facilitator. A good facilitator can handle disruption, and the facilitator may consider the person to be far more manageable than would persons who are not trained and experienced in facilitation.

Ironically, in the case of facilities in particular, some of the stakeholders who may be the most critical can be the hardest to recruit. Some environmental organizations simply do not wish to appear to be aligning themselves with industry or government by participating in a CAP. In some cases, activists may also resist coming to a facility for tours or meetings out of concern for their health or safety. Sometimes these people can be brought to the table later in the CAP's existence, after other CAP members persuade them of the legitimacy of the dialogue. Another approach is to invite such people to visit a CAP meeting at an acceptable location to air their views, at least on a one-time basis. This will serve to enter their concerns into the CAP's dialogue, even if the person in question still prefers not to become a formal member.

As alluded to in the preceding paragraph, some CAPs meet on the premises of the facilities they are set up to discuss. Meeting location is a decision that will depend on the situation. For example, if prospective CAP members express concern about coming to a facility, then another location may be chosen, at least for the initial meetings. Many CAPs meet at public facilities, such as library or community center meeting rooms, or at restaurants where a meal can be served as part of the meeting. Some CAPs also hold open meetings that can be attended by any interested member of the public. If there is a great deal of interest in the facility or site, this is a good way of extending the reach of the dialogue and the educational component of the meetings beyond the immediate CAP members. Holding open meetings can also be an effective way of involving persons who, as discussed in the preceding paragraph, don't wish to join the CAP, but who may at least wish to monitor the discussion. Some CAPs make use of a hybrid approach, holding closed meetings most of the time and then holding an open meeting or making a "report to the community" once a year. In such cases, facility or project personnel or other persons connected with the site or facility may give presentations. CAP members can join in during a question-and-answer period, by either asking questions or by providing answers or comments on what they have learned. The length of CAP meetings, the time of day, and the day of the week of the meetings should be set based on the schedules of the persons whom organizers of the CAP wish to attract.

Frequency of CAP meetings varies substantially. The key factor in meeting frequency tends to be the level of concern about the site or facility that the CAP was formed to discuss. For example, if community members—and by extension, members of the CAP—are extremely concerned about a site investigation and its findings, monthly meetings may be in order. The initial meetings may involve a great deal of "catch-up" as CAP members learn about the history of the site and the process by which site investigations are conducted. Meeting frequency may drop back to bimonthly or quarterly—or they may be held in conjunction with the completion of certain phases of an investigation or cleanup—once CAP members feel that they have a handle on what is going on.

Some CAPs meet on an as-needed basis, or hold special meetings to address issues as they arise. For example, if a facility with a CAP receives notification that it is about to be fined or cited for a violation, it may call a CAP meeting to inform the members before the news is publicized, either to provide an explanation and to give CAP members the opportunity to ask questions or to ask for suggestions on how facility management should communicate with the rest of the community about the impending news.

CAP meetings may be facilitated by an outside facilitator, such as a consultant or a member of a civic group, such as the League of Women Voters, or by a person chosen by the CAP from among its membership (i.e., a chairperson). The facilitator typically sets the agenda with input and suggestions from the CAP, encourages full exploration of topics and input from all CAP members, and acts to defuse destructive behavior to prevent harm to the group process. Some CAPs may be facilitated or chaired by a person associated with the site or facility that formed the CAP, such as a facility manager. It should be noted that this approach is probably not the best choice and can compromise the dialogue, since it may be difficult for someone with a vested interest in a facility or investigation to objectively facilitate discussion that is critical or angry in nature.

Setting Goals and Objectives for the CAP

Recruiting appropriate stakeholders, in this day of tight schedules and limited free time, can be difficult. Persons responsible for setting up the CAP will often need to "sell" the overall idea of the CAP and its benefits to the community. Initial goals and objectives should be identified to aid in explaining to prospective members what the CAP is all about. These preliminary goals and objectives should be based on the issues uncovered by the community assessment and by the situation or pending issues at the site or facility. For example, if facility management is planning to begin a major expansion or permitting activity in the next year or so, then the goals and objectives need to consider the issues that will surround such activities. Once the CAP has been formed and is operating, it can then establish additional goals and objectives of its own.

Persons responsible for setting up a CAP must consider, at the outset, how they are going to deal with suggestions from the CAP and ensure that expectations are set accordingly from the start. For example, it would not be wise to say to a facility CAP that management intends to implement (or try to implement) all of the suggestions the CAP puts forth, since many suggestions may not be implementable given technical and business constraints.

Some other CAPs actually do fulfill an advisory role, however, and their suggestions are intended to be implemented. In such cases, care should be taken to ensure that the goals and objectives or charter of the group spells out exactly what issues the CAP is supposed to explore so that feelings and egos won't be hurt by the CAP straying into areas that are not meant to be subject to the CAP's advice.

Life Span of a CAP

Some CAPs are meant to have a limited life span—they address a specific issue, such as a site undergoing investigation and cleanup (although this can be a number of years at many large or complex sites). Other CAPs, such as those formed by facilities, are often intended to be permanent fixtures to ensure that channels of communication remain open. Maintaining interest in CAP participation can be difficult if there are no pressing issues. One technique used to maintain interest in a CAP is to set time limits for CAP member participation. New members are brought on board as other members' terms expire. This technique can work well in communities that have a large pool of interested stakeholders to draw from. Another technique for keeping interest is to turn the CAP's attention to other community issues, or involve other facilities in the area. This removes the group's attention from the originating facility—but it does at least keep a line of communication open in the event that an issue with the facility is raised in the future.

Interest in CAP involvement may also die out, despite continued community concern, in cases in which CAP members perceive that their concerns are not being listened to, or that no improvements to the site, facility, or project operations are being planned or implemented. This is a very serious situation, which can be viewed as public evidence that the community relations dialogue is not working and that the parties in charge of the facility are not interested in public concerns. Reviving a meaningful dialogue and re-establishing trust after such a breakdown can be difficult. The remedy to such a situation is to avoid it from the start through clear communication about what is and isn't possible, and by following through with those actions needed to fix serious situations. It does no good for a facility to agree that it shouldn't have as many spills as it does if the facility management isn't willing to take steps to rectify conditions that lead to spills. Eventually, stakeholders' attempts at a dialogue will simply turn to accusations and disappointment.

Size of the CAP

CAPs can vary in size from five to 25 or more. The size will depend on the number of stakeholders that may be interested in or affected by the site or facility. A CAP for a small site or small facility may include a few neighbors and one or two municipal officials. A CAP to address a major site or large facility that could have an impact on air and surface or groundwater over an extended area may need to include members from several communities and a number of stakeholder groups. Although a CAP's membership should represent the community, groups of more than 15 can become unwieldy to facilitate.

Using Focus Groups as an Alternative or Supplement to CAPs

Although facilities tend to gravitate toward CAPs in keeping with their long-term presence in a community, focus groups can be a viable alternative for facilities that aren't sure that they wish to commit to a CAP. Thus, a focus group can serve as a middle ground between the formal input of a CAP and relying on other, less formal and less direct means for obtaining input from stakeholders. Focus groups can also be used as a supplement to CAPs if certain stakeholders—such as activists—do not wish to be affiliated with the CAP. In such cases, these stakeholders may be willing to sit down for a focus group session—basically a forum for them to express their concerns and opinions while still keeping their distance from facility management that they may not trust. The key here will be to have a neutral facilitator, or a facilitator who is sympathetic to their views, and to conduct the meeting in a neutral location.

Focus groups can be set up two different ways. As we discussed in chapter 3, focus groups can be composed of members of a single stakeholder group or similar types of groups (e.g., a focus group composed of persons from different environmental organizations that may have some differences in their agendas, but are essentially united in their interest in environmental protection) or they can reflect a cross section of the community. If the focus group is intended to represent a cross section of interests in the community, then the same care used to recruit CAP members should be taken to ensure that the feedback they provide reflects the views of the community.

Occasionally, focus groups may be used in conjunction with CAPs, since CAPs may not be able to provide accurate feedback on certain issues because of the level of knowledge its members have obtained through CAP participation.

CHAPTER

5

COMMUNITY RELATIONS PROGRAM PLANNING

This chapter discusses the community relations program planning process. It integrates the information from the preceding chapters to describe how a community relations plan should be prepared, and includes several worksheets that persons charged with designing a community relations program can use. Additionally, this chapter explores methods for evaluating the effectiveness of a community relations program.

A community relations plan should be a "living document" that is amended as necessary to maintain a good fit between the community relations activities and their timing and frequency, and the concerns, wants, and needs of the community. In addition to identifying the stakeholders who should be involved and their information needs, the plan should provide timetables for activities and identify the resources that will be required to meet the goals and objectives of the community relations program.

PLANNING FOR COMMUNITY RELATIONS PROGRAMS

An effective community relations program requires that a number of actions take place. These actions include:

- Identifying all stakeholders and considering their wants, needs, and issues.
- Identifying appropriate channels of communication for reaching stakeholders.
- Promoting a dialogue that allows for the free flow of information, including:
  - Keeping stakeholders informed of developments at the facility, site, or project;
  - Providing understandable information on the topics of interest or concern to stakeholders; and
  - Actively soliciting and encouraging feedback, suggestions, questions, and concerns from stakeholders.
- Exploring options that will accommodate community concerns or incorporate stakeholder suggestions.
- Following through on commitments, when made, to remedy or modify a situation.
- Showing that the commitment to protect human health and the environment is real through superior environmental performance.

We discussed the process of identifying potential stakeholders and their wants, needs, and issues in chapter 2. After the assessment, the community relations planning effort shifts to determining:

- Which groups or individuals appear to be stakeholders in regard to the issue in question and should be included in the community relations dialogue.
- The types of information stakeholders want or need to receive, based on their requests, misunderstandings, or lack of knowledge, as well as the information they will need to receive in order to answer questions that are likely to arise as they learn more about a facility, site, or project, or as work progresses on a new phase of an investigation, cleanup, expansion, and so forth.
- The best channels of communication and activities to conduct the community relations dialogue.
- What additional actions need to be taken or problems need to be addressed to further the community relations effort.

Plans for community relations programs may be prepared either in response to an issue (reactive) or before an issue is generally known in the community (proactive). As we stated in chapter 3, the latter approach—the proactive approach—is superior because it provides managers with substantially more control over how and when information will be released. As we have noted, proactive disclosure is often received in a more positive manner than information provided in reaction to community concerns or opposition.

Community relations plans should be "living documents." Unlike assessment reports, which basically provide a snapshot in time (though they may be updated periodically), the community relations plan needs to fit the circumstances (see Figure 5.1). Thus, as an issue and the dialogue surrounding it unfold, the community relations program may need to change appreciably to reflect new stakeholders, the strengthening of relationships, or the desire for additional information on new issues or evolving aspects of the original issue. The plan should be updated accordingly, as well.

A community relations plan should be more than simply a compilation of activities. It should provide a discussion of why certain activities are chosen, what their preferred timing should be, and how the activities will help reach the stated goals and objectives of the plan. Creating such a document helps the community relations specialist ascertain whether the plan, as designed, fits the community and the issue(s) and will further the goals and objectives of the community relations effort.

The community relations plan should also identify the resources that will be necessary to implement and maintain the community relations effort. Resources include written materials, such as fact sheets and questions and

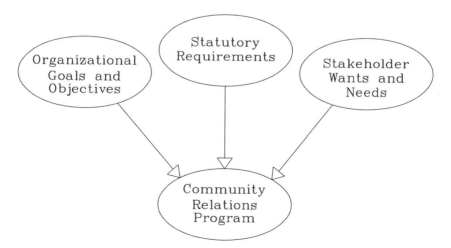

■ **FIGURE 5.1 Plans for Community Relations Programs Must Consider a Number of Variables.**

answers, displays for open houses or exhibits, time for the community relations specialist and other facility, site, or project personnel who will be involved, time or availability of "outside experts," and other materials or services that will be needed. A model outline for community relations plans appears in the box below.

Most of the information in this chapter is aimed at community relations programs designed to address fairly serious and/or complex issues. There are many other instances in which only a handful of people may be interested in or affected by an environmental issue, or the issue itself may be relatively simple and can be easily explained. In such cases, a written community relations plan isn't necessary. Still, the information in this chapter can provide important guidance to community relations specialists and others who must deal with these simpler scenarios, even when written plans aren't prepared.

USING THE COMMUNITY ASSESSMENT TO IDENTIFY COMMUNITY NEEDS

The community assessment provides the majority of the information that will be needed to design an effective community relations program. As described in chapter 3, additional information may also be gathered via questionnaires, focus groups or small group meetings, or additional one-on-one interviews in very serious or complex situations involving diverse and/or populous communities.

Annotated Outline for the Community Relations Plan

I. Executive Summary
Summarizes reason why community relations program is being implemented as well as community concerns and questions or potential concerns and questions regarding an issue. Also provides community context surrounding an environmental issue or potential issue. Provides brief overview of the community relations program, including activities, key messages that need to be conveyed, and primary stakeholders who are targeted.

II. Goals of the Community Relations Plan
Lists goals that the organization wishes to accomplish through the community relations plan (e.g., obtain public support for permit application; minimize concerns about/gain support for remediation of site; gain public input into decision-making process re-

garding harvesting or extraction of natural resources). Also provides brief discussion of why achieving these goals is important.

III. **Situation Analysis**

Describes reason why community relations program is being implemented and provides background information on the concerns, questions, community dynamics, and statutory requirements (if any) that are driving the program or that need to be considered in its design.

A. **Description of Facility, Site, or Project**

Describes the facility, site, or project that is the focus of the environmental issue in question and the community relations effort. Description should include the history of the facility, site, or project relevant to the issues affecting/likely to affect it as well as community relations activities that have been undertaken or significant relationships that have been established.

B. **Overview of Community**

Provides brief overview of community dynamics, geographic extent (for purposes of the specific issue in question), history, other issues that may influence attitudes or opinions about the environmental issue in question and how stakeholder groups fit within the structure of the community.

C. **Issues Affecting Facility, Site, or Project**

Provides a detailed overview of the environmental and nonenvironmental issues affecting the facility, site, or project and why a community relations program is being implemented to address them.

D. **Upcoming Activities Requiring Disclosure or Public Involvement**

Lists any statutory requirements for disclosure of information or for public involvement or other impending activities, such as an expansion of a facility or extractive operations, that may draw public attention.

IV. **Description of Stakeholders**

Provides detailed information on stakeholders' agendas, goals and objectives, level of understanding, areas of concern, preferred channels of communication and activities, and other stakeholder groups with which they are allied.

V. **Community Relations Plan**

Provides detailed description of the community relations program and how its elements meet the stated goals and objectives. Also provides a timetable for the program and lists the resources that will be required to carry it out.

A. Activities, Communication Vehicles, and Channels of Communication

Describes activities, communication vehicles, channels of communication, key messages, or concepts that will be conveyed, and the type of input or feedback that is desired. Also indicates which stakeholder groups will be targeted by each activity and communication vehicle and why they will be targeted.

B. Timetable for Community Relations Program

Provides critical dates and logical sequence of events, preparation/presentation of information.

C. Resource Requirements and Budget

Lists resources that will be required to implement community relations program.

Which Groups or Individuals Should Be Considered Stakeholders?

In chapter 2, we stressed the importance of identifying all potential stakeholders during the assessment process, and leaving the decision as to which ones should actually be included in the community relations dialogue until the planning stage. This practice minimizes the likelihood that groups or individuals that may be affected by, or interested in, the issue will be left out.

From a practical standpoint, however, decisions do have to be made about who the stakeholders really are in regard to a specific issue so that resources aren't wasted reaching out to those who aren't interested. This decision will lie somewhere between excluding everyone but the next-door neighbors and including the entire world. A good rule of thumb is to include all those who are affected, perceive they are affected, or want to be involved. This still leaves a lot of open territory for those who must make budgetary and logistical decisions regarding a community relations program. The following paragraphs provide some suggestions on how a community relations specialist might go about determining who should be involved and to what extent.

It should be noted that in cases in which the "public" issue includes "private" issues between, for example, an organization and a neighbor, managers' judgment, legal considerations, and the neighbor's interests and preferences will limit the sharing of "private" information with other stakeholders. The sharing of information will also typically be limited in situations involving enforcement actions or other legal issues. The "public" component of the information would be shared with all stakeholders included in the dialogue, however.

Primary and Secondary Stakeholders   When designing a community relations program for issues that are likely to involve more than a handful of people, it can be useful to divide stakeholders into primary and secondary categories. Primary stakeholders would be regarded as persons or groups that have a direct connection or direct, articulated interest in the facility, site, or project in question, and, as a result, are likely to have an interest in participating in the actual dialogue surrounding the issue. Primary stakeholders would include the following types of persons or groups:

- Those who are directly involved because of their proximity to a facility, site, or project, such as neighbors.
- Those who have an interest in the issue because of the positions they hold, such as elected and appointed officials and other local government staff.
- Those whose "level of interest" suggests that they would likely want to be part of the dialogue concerning the issue. The "level of interest" category may include persons or groups involved in environmental or community organizations, other persons or groups that have actively expressed interest or concern, or persons or groups that have some other connection, such as employment at or near the facility, site, or project in question. "Level of interest" may be intentional, as in the case of environmentalists, or it may be incidental, as in the case of employees. In either case, such individuals or groups have a direct connection to the facility, site, or project.

As we will discuss below, "outsiders" who have been invited into (or inserted themselves into) the situation also typically function as primary stakeholders.

Secondary stakeholders would be others who may have an interest in an issue, but who are unlikely to expect to take part in the actual community relations dialogue. For example, secondary stakeholders may include people who live at some distance from a facility, site, or project. They may have some concerns about the issue in question, and be following stories about it in the media because it is unfolding in their town or county, but they do not perceive themselves as personally affected.

Other secondary stakeholders can include local businesses or business groups that are watching how facility, site, or project management is behaving or being treated and may have concerns about how the issue in question could reflect on other area businesses. Thus, the secondary stakeholder category is basically a catchall for all of those people or groups that are "following" a situation.

Although the community relations program will place much of its emphasis on working with the primary stakeholders—those who are taking part in the actual dialogue—the secondary stakeholders should also be kept in the community relations loop. For example, information about the issues surrounding a facility, site, or project may be submitted to the local media to generate news stories to inform both primary and secondary stakeholders about activities, conditions, or new developments. Advertisements can also be used to fulfill this function and to publicize public meetings and the availability of information in repositories or via the management of the facility, site, or project. In some cases, fact sheets or other information may also be mailed to whole communities or portions of communities, targeting both primary and secondary stakeholders.

There are two reasons why attention should be paid to secondary stakeholders. First, providing people and groups that have at least some level of interest in the issue with information should mitigate their concerns that the issue isn't being addressed—a perception that could lead to their intervention in the dialogue with the primary stakeholders or the expression of negative statements based on misinformation or lack of information that could adversely influence primary stakeholders who are involved in the dialogue.

Second, it ensures that, if the community relations specialist has misread the community or the issue and failed to include all of the relevant primary stakeholders in the actual dialogue, then these groups and individuals are likely to have been included in the secondary stakeholder category. These miscategorized stakeholders are more likely, then, to either move themselves or be moved by the community relations specialist from secondary to primary status, and are less likely to perceive that they were intentionally excluded—with all of the anger that such perceptions can bring with them. The door should be open for secondary stakeholders or new arrivals to become more involved in the community relations process.

Stakeholders may also move from secondary to primary stakeholders as an issue evolves. For example, as the dialogue about a facility evolves from discussion about wastewater discharges to discussion of air emissions and risk management plans for the accidental release of toxic chemicals, the number of stakeholders who may perceive themselves as potentially affected on a personal level can increase. As we discussed in chapter 2, the geographic extent of concern often varies according to the type of issue. Thus, facilities that operate for many years and some site cleanups or projects of long duration may see significant shifts in both the amount of concern and the number and extent of stakeholder groups that would consider themselves primary stakeholders (e.g., as personally affected in some way).

By building methods for reaching secondary stakeholders into a community relations program, both the "learning curve" about the issue and potential hostility from the perception of being "kept in the dark" can be minimized as secondary stakeholders convert into primary stakeholders.

In one case, the investigation of a contaminated site went along fairly smoothly, with community relations efforts concentrating on the near neighbors and municipal and county officials. Additional stakeholders clamored to join the dialogue when the remedial alternatives—which included on-site incineration—were being discussed. Groups that were opposed to thermal treatment, as well as other community residents some distance from the site (*and* the near neighbors), became concerned when they learned that an on-site incinerator was being considered to treat the contaminated soil.

Involving Leaders, Neighbors, and the "General Public"

The question often arises as to what is meant by "stakeholder involvement." Does it mean *everyone* in a stakeholder group? Or does involvement by representatives of these groups constitute "stakeholder involvement"? More often than not, group participation in the actual dialogue will, in fact, involve only a handful of representatives.

Communities, and the groups of people that comprise them, invariably include formal and informal leaders. Formal leaders include elected officials and those who hold specific titles, such as fire chief, as well as recognized leaders of civic organizations, churches, environmental organizations, and the like. Informal leaders include people who don't hold a title per se, but whom others in the community look to, often because they are considered knowledgeable. For community relations purposes, they may be people who tend to be spokespersons for neighborhoods or who have an interest in environmental issues, although they aren't presidents of local environmental organizations.

With the exception of "near neighbors"—those people who are directly affected by a facility, site, or project, and, thus, are involved or who have the potential to become involved out of self-interest—many community relations activities are aimed primarily at communicating with formal and informal leaders as representatives of stakeholder groups. Activities such as workshops, small group meetings, tours, and availability sessions are typically aimed at the leaders, as well as the near neighbors and other people who, for whatever reason, have expressed a keen interest in the issue. This makes sense in many communities, since, given most people's busy schedules, the leaders are the people most likely to make time to engage in a dialogue about the issue in question. Additionally, as leaders, they are expected to act as conduits of information for other members of the community, both by sharing the knowledge they obtain through the community relations activities with other community residents and by conveying residents' concerns and questions during the dialogue.

It must be noted, however, that attention must be paid, during the design of the community relations program, to the role that formal and informal leaders play in the specific community to ensure that all of those individuals who are affected will be well represented by "leaders." Is the dynamic of the community such that including only "recognized" formal and informal leaders will still leave some potential stakeholder groups unrepresented? Will extra effort be necessary to ensure that some demographic groups, whether identified according to race, ethnicity, income, level of educational attainment, or length of time in the community are indeed represented in the dialogue?

Community relations specialists will need to keep track of who is participating in community relations activities throughout the duration of the program to ensure that groups or individuals who do have an interest, or are affected by an issue, don't feel that they aren't being adequately represented in the dialogue. If they aren't being adequately represented, then efforts may need to be taken to reach out to them more aggressively and to find individuals who can function as "leaders."

Although leaders (and near neighbors) are the most common and, typically, the most active participants in community relations dialogues, this is not to say that members of the general public should be excluded from community relations activities or from the communication effort. In cases of serious concern, many members of the "general public" may wish to be personally involved by attending meetings or open houses or requesting fact sheets. Fact sheets, news stories, exhibits, open houses, large and small group meetings, and many other activities or communication vehicles are aimed at the general public as well as at leaders. In the environmental community relations arena, not only can new "leaders" emerge at any time, but public sentiment about the environmental issues in question can also influence the "leaders" that are being targeted by the community relations program.

Dealing with Disruptive Stakeholders

It is rarely wise for the organization responsible for community relations in support of a facility, site, or project to attempt to exclude disagreeable stakeholders from a community relations dialogue. There are two reasons for this. First, if persons or groups have an interest in or concern about an issue that falls under the purview of the community relations effort, they have a right to be involved in the dialogue[1]. Disruptive behavior can often be dealt with effectively through good group processes, such as effective facilitation, or other means of managing the dialogue. Second, attempts by the organization responsible for the community relations program to expel or exclude difficult stakeholders—even if other community residents agree they are difficult to deal with—can lead to the perception that the organization is exerting undue control over the community relations dialogue. Thus, it is best to attempt to work around disruptive stakeholders and allow other community residents to tell them that they are out of line. Excluding these people or groups should be a last resort—and may not be possible in the case of statutory requirements or issues involving public sector organizations.

Making Decisions Regarding "Outside" Stakeholders

One of the toughest decisions to make in regard to stakeholders is to determine whether, or to what degree, groups or individuals that are clearly from outside the community boundaries should be involved in the community relations dialogue. For example, should a group that monitors the water quality of a lake 30 miles away be considered a potential stakeholder—either primary or secondary—in regard to a situation involving the water quality of a local lake? How far away is too far away? One would have to ascertain through some research whether the group would actually be *interested* in taking an active role in the dialogue concerning water quality in a lake located in a distant community. Because of the distance, it is likely that such a group

---

1. Again, as we have stated, certain "private" aspects of an issue (for example, certain specific issues between the organization and neighbors where privacy or legal issues are considerations) may not be shared with all stakeholders involved in the dialogue.

either wouldn't be considered a stakeholder at all, or that it would be considered, at best, a secondary stakeholder.

Outsiders may also elect to become involved in an environmental issue on their own or at the invitation of local stakeholders. This type of outsider involvement is often viewed with exasperation by managers of facilities, projects, or sites, since outsider involvement typically broadens the scope of the questions and concerns that "real" community residents have raised. Managers often view these people as interlopers who have no legitimate claim on local issues. Whether this is true or not, outsiders who enter the process, either on their own or at the invitation of local stakeholders, must be dealt with within the community relations process.

Another scenario can occur when an outside group wants to be involved, and local residents—or some local residents—prefer that they not. Should these outsiders be considered stakeholders? For example, some environmental activists will "come into town" to organize opposition or at the behest of local residents.

Initially, at least, it is best that the organization responsible for the community relations not take action itself to exclude such parties—particularly if some local residents indicate that they want the outsiders involved. These outsiders should be treated as primary stakeholders. This *does not* mean that misinformation being spread by such groups should not be countered—it should be. However, countering misinformation, even forcefully, is well within the purview of the community relations process and demonstrates a willingness to work toward the resolution of relevant issues with the community.

Why not simply exclude outsiders? As we explained previously in regard to local, disruptive stakeholders, attempting to exclude outside activists could be construed as an effort to exert undue control over the dialogue or to keep out those who may possess "superior knowledge" about environmental issues. Thus, it is better to concentrate on building strong relationships with local primary and secondary stakeholders and let them take their own steps if the outsiders appear to be disrupting the dialogue in a negative way. Local stakeholders may be willing to listen to outsider activists, but they will also disregard their comments or forcefully ask them to leave if their activities appear to be impinging on local interests. If most stakeholders want outsiders to leave and a minority wants them to stay, again, it is likely that their effectiveness within the dialogue will be minimized simply because most stakeholders aren't listening to them. Still, even in such situations, it is best if action against outsiders (even unpopular ones) comes from other stake-holders, *not* from the organization responsible for the community relations program.

It should be noted that outside activists can play a beneficial role in the community relations process. Activists who are knowledgeable about environmental issues can often confirm whether the information provided by a company or governmental unit is correct—and their pronouncements can do a lot to promote the credibility of the information. The likelihood that such individuals or groups will play a positive role is typically governed by such factors as whether they are seeking to use an environmental issue for fundraising purposes, as well as their general stance toward government or industry. Often, groups or individuals that are in the community to raise money or are trying to prove a point by "punishing" a company or government entity will leave on their own if a good community relations program has been initiated because they cannot make substantial inroads into the community. Activists who genuinely want to work to resolve the issue are more likely to remain and contribute in a positive fashion to the dialogue. More information about working with community-based activists appears in chapter 6.

Dealing with "Controlling" Stakeholders

As we discussed in chapter 2, some formal or informal leaders or segments of a community may attempt to control who is involved in the community relations dialogue. Their motivations may be benign—they don't believe that residential neighbors know enough, or they believe that the interests of these people are already being represented by other groups—or they may reflect a desire to ensure that they retain control over a community issue. It is important in community relations not to allow the interests of some to ride roughshod over the interests of others, such as neighbors, who have a very legitimate claim to involvement in the community relations dialogue. Specific community dynamics regarding decision making and power sharing all come into play when dominant leaders or segments seek to control the community relations dialogue.

Although it is important, for example, to work with elected officials and other formal and informal leaders and not undermine their positions in the community, community relations specialists must emphasize the need to reach all stakeholders and include them in the dialogue. This can be tricky, and can require community relations specialists to be quite creative to achieve the desired results without alienating controlling leaders. For example, some situations may require working through "formal" leaders to make initial contacts with stakeholders, or religiously reporting back to the "formal" leaders about the general status of the community relations effort. Ultimately, however, community relations specialists need to establish trust, relationships, and a positive dialogue (which may include keeping some comments in confidence) with stakeholder groups separate from an overly controlling community power structure. Doing this often involves quietly reiterating the need to work with all stakeholders, and then going ahead and doing so.

In one case involving controlling leaders, the mayor and city council of a community were highly suspicious of attempts by a local facility to establish a community advisory panel (CAP) that would include local residents—including some who were not allied with the ruling political party. This was definitely a power issue in this politically embattled community. The officials insisted that all communication should go through them and, in fact, had objected to the performance of a community assessment. (The assessment showed serious concerns about the facility on the part of the near neighbors and some other opinion leaders.)

The community relations specialist managed to get the mayor and city council to agree not to actively oppose the CAP by reiterating a desire to work with the elected government and by clearly communicating the corporation's need to establish the CAP to improve relations with the community.

In this case, the answer was one part diplomacy and one part clear communication of corporate needs. The diplomacy involved including a city council member on the panel (an action that the facility's management was planning to take anyway) and emphasizing facility management's appreciation of the importance of working with the city council. The communication of corporate needs basically involved educating the mayor and the council members about the company's emphasis on good relations with the community and with residential neighbors and senior management's commitment to working with all of its neighbors.

Facility management also shared with the mayor and city council information about the types of people it was planning to have on the panel, and why these people were important (omitting comments about the need to include representatives of the rival political party, which was gaining strength). As politicians, the mayor and council members could see the logic in the panel, and made a few excellent suggestions that tracked well with what the facility managers and the community relations specialist were planning to do anyway. It is important to note that facility management did not "buy" agreement by "watering down" the composition of the panel.

Working with Individual Stakeholders and All Stakeholders Together

The question that should be raised next is the extent to which stakeholder groups should be dealt with as discrete units or combined for purposes of activities or communication. On the one hand, the interests of individual groups need to be weighed; groups that have special, legitimate interests shouldn't have their voices drowned out by the others.

On the other hand, treating each stakeholder group as a separate unit can create significant problems. For example, resolution of a specific environmental issue sometimes requires that stakeholder groups within a community resolve their own differences of opinion on the issue, as well as addressing the concerns they may have about the facility, site, or project in question. From a practical perspective, stakeholder groups often have many of the same concerns; thus, communicating separately with them is a waste of resources. Additionally, communicating with discrete groups can give rise to accusations that managers of a facility, site, or project are giving different information to different people. In fact, this does happen, since, invariably, different questions will come up and be answered when the communication is segmented through a number of one-on-one or small group meetings. As we've stated, one of the recommended strategies for addressing stakeholders' concerns is to use a number of one-on-one meetings and small group meetings to allow for a more intimate dialogue. This *is* highly desirable; however, segmenting stakeholders so that virtually all of the communication involves only single groups is *not* wise.

So, how does one accommodate differences among stakeholders and their interests? The best alternative is to use a combination of "general" information or events that are aimed at everyone (with translations into a second language, when applicable) along with some small group or one-on-one work both to move the dialogue along by allowing more involvement of all parties and to address the concerns or issues of specific stakeholder groups.

Limitations on Stakeholder Input

Some primary stakeholder groups may need to stand a bit off to the side during the dialogue because of the positions they hold in government or in the community. For example, an environmental agency or local government unit may be very much a primary stakeholder, and would definitely be part of the dialogue about an issue, but its personnel may have to stand back from airing certain views in order to retain its legitimacy as a public entity. In such cases, agency or local government personnel may have opinions about the types of comments that neighbors or activists are making; to avoid the appearance that they are "siding" with owners or others responsible for a facility, site, or project, however, they must refrain from volunteering these views or risk impairing their own reputations as protectors of the community's interest.

Thus, it is the organization responsible for the facility, site, or project that must make appropriate information available and ensure that stakeholders understand it. Local government officials or agency personnel can then confirm, if asked, that the information provided is valid. This function is considerably different (and often more powerful, in terms of influencing public opinion) from asking a government entity (or some other important stakeholder) to stand arm in arm with the organization that it is supposed to be regulating or overseeing.

Of course, there are other times when agencies or local officials do stand up and make statements that are positive in regard to the organization responsible for the environmental issue in question. Such pronouncements are typically made on behalf of the *public*, however—an important distinction that community relations specialists need to consider when designing community relations programs and identifying likely supporters or persons or groups that they can work with to help educate other stakeholders. Thus, it is important to consider carefully the role that certain stakeholders can or are willing to play when designing a community relations program.

Identifying Stakeholders' Information Needs and Wants

As we discussed in chapter 3, fine-tuning the communication effort may require some additional research, such as the use of questionnaires, focus group or small group meetings, or one-on-one interviews, to home in on specific issues as well as to test messages and determine how best to present information to stakeholder groups. This is information that needs to be carefully spelled out in the community relations plan. The level of detail that different stakeholders might want should also be ascertained.

It is often possible to identify ahead of time some of the topics that are going to be of interest to stakeholders, either as they learn more about a topic or as an issue develops. For example, many of the questions stakeholders will have about the investigation and cleanup of a contaminated site will track along with the progression of the investigation and remediation process (see Figure 5.2). Following are some common questions that are likely to come up in the case of an investigation.

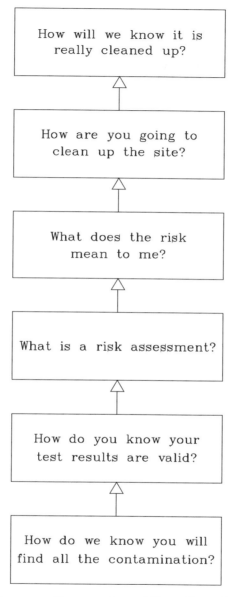

■ **FIGURE 5.2 Common Progression of Questions Regarding the Investigation of a Contaminated Site.**

Common Questions About an Investigation

Discovery of Contamination:

- What did they find?
- How did it get there?
- Who is going to take care of it?
- Am I at risk?
- What does this mean?

Investigation:

- What are they testing for?
- Are they investigating the right areas?
- Will the test results really show what's there? Could the results be manipulated? Who approves them?
- Does the investigation put us at risk?
- Who decides if the investigation is being conducted correctly?

Reporting of Investigative Results:

- What do these results mean?
- They found something—are we at risk?
- They say we aren't at risk—how do we know they are telling us the truth?
- Has anyone else looked at what they did and figured out if they did it right?

Cleanup of Site:

- How are they going to clean it up?
- Why did they choose that method?
- Are we at risk from the cleanup?
- How do we know they are getting everything?
- What is going to happen to the property now?
- How long will it take?
- How clean is clean?

Overriding all of these questions about investigations and cleanups are other concerns that may have to be dealt with throughout the process. An example of these include:

- Is it *really* safe to live here?
- Why does it take so long to clean it up?
- What will happen to my property's value?
- Are they telling us everything?
- Is anyone overseeing all of this to be sure it's being done right?

Community relations plans that anticipate the logical flow of the concerns and questions that are likely to accompany an issue are typically more effective than those that do not, and are certainly more likely to embody a proactive approach. As we will discuss later in this chapter, many facilities, sites, or projects are subject to statutory requirements for disclosure of information or for public involvement. The plan preparation process should attempt to anticipate questions or concerns relative to the information released under these statutory requirements.

Identifying Channels of Communication and Community Relations Activities

Chapters 2 and 3 have touched on the importance of identifying appropriate channels of communication and community relations activities to match the preferences and needs of community residents. Channels of communication and community relations activities must also be chosen with the specific issue in mind.

In some cases of limited scope, for example, when only a few people may be affected by or interested in an issue, relatively few contacts may be necessary to provide stakeholders with adequate information and/or gain their input. One-on-one meetings and/or a fact sheet, and a few follow-up calls, may be sufficient. In more complicated situations—for example, those that affect more people, are perceived as posing significant risk or impact on people's lives, property, or the environment, and/or are likely to be long-standing in duration—community relations specialists typically like to employ a variety of activities and channels of communication to provide information and promote a dialogue.

As a rule of thumb, the authors suggest that community relations programs in support of complex or serious situations include a mix of the following:

- One-on-one conversations (neighbor visits) and small group meetings with neighbors and other stakeholders having significant interest in the issue.
- Availability sessions and workshops for neighbors and formal and informal leaders, such as local officials and neighborhood, community, social group, or environmental group leaders.
- Written materials, such as fact sheets, for wide distribution.
- News stories and advertisements for distribution of information to the general public in the community.

■ Information repositories containing all of the information that has been made available on the facility, site, or project, as well as other background and publicly available information for anyone who is interested in reviewing detailed information.

■ Community advisory panels (CAPs) should be considered for facilities or projects of long duration (including some site investigations and cleanups).

As we discussed in chapter 4, many other activities can also be used, including presentations to local organizations, open houses and tours, postings of information in areas of the community where they will be seen by persons who don't typically read the local newspaper, and so forth. These other activities can be extremely effective and appropriate, and can be added to the mix depending on the issue in question. Environmental issues that involve technical information (which includes the majority of serious environmental issues) typically require a combination of written materials and presentations to give community residents a good understanding of the issue in question.

Do Not Underestimate Communication Needs One of the sources of information community relations specialists use to determine which channels of communication or activities should be used in a community relations program is suggestions from people who are interviewed during community assessments or who participate in focus groups. The authors have found that, although these people often have very good ideas, they consistently *underestimate* the amount of effort it takes to communicate successfully about complex or contentious issues. For example, when community relations specialists are implementing dialogues over issues that are not yet generally known in the community, or in cases in which residents are already strongly opposed to certain courses of action, such as the permitting of a hazardous waste incinerator, it will often be necessary to do a lot more in the way of communication than persons interviewed during an assessment or focus group may suggest.

Persons interviewed during an assessment may suggest newspaper stories as the best means for reaching stakeholders. Since this is their preference, newspaper stories would definitely be one channel of communication; a newspaper story is unlikely to convey sufficient information to promote a constructive dialogue, however, especially if the issue is complex or the subject of significant opposition. Thus, the objectives of the newspaper stories would include providing background information and publicizing other community relations activities, such as small group meetings with stakeholders, that would provide more detailed information. This would serve to let community residents know that the issue requires more attention than they originally might have thought, and provide information on how they can learn more and become involved in the dialogue.

Another example that will likely require more communication than might be apparent to some interviewees involves disclosure of a facility's risk management plan and worst-case accidental release scenario under Section 112(r) of the Clean Air Act. Under these regulations, this information will be made available to the public; however, in many cases, stakeholders will know little or nothing about the facility's operations or local emergency response plans or capabilities.

Thus, educating community residents about what the worst-case scenario means, providing other information (e.g., information on local emergency planning and operations), and receiving stakeholder input, will probably take substantial effort on the part of facility management and local emergency response personnel. This means that multiple contacts, through multiple channels of communication, will likely be necessary to ensure stakeholder understanding.

Other Issues That Need to Be Addressed

As we have discussed, resolution of environmental issues can involve more than communication. In some cases, it may be necessary for the management of a facility, site, or project to take additional steps beyond the dialogue with stakeholders to resolve community concerns. For example, if concerns voiced during a community assessment or focus group indicate that emergency response capabilities are sorely lacking, then the community relations plan should indicate this. Facility management may then elect to begin a proactive dialogue with local emergency responders to identify areas that can be improved as part of the community relations effort, without waiting for further dialogue with community residents to suggest it. In other words, the assessment can often point up actions that can or should be taken or improvements that can or should be made without having to wait for additional input from area residents. These recommendations can be built into the community relations plan. Making such moves in a proactive fashion is a good way to get a dialogue with stakeholders off on the right foot.

It is also important to identify those problems over which management of a facility, site, or project has little or no control, but that are likely to play a role in the community relations effort. For example, a political battle over the development of property adjacent to a facility or site can have a major impact on the opinions of certain stakeholders about activities or problems related to the facility or site, and must be taken into consideration as the community relations plan is developed.

In one case, property adjacent to a RCRA permitted hazardous waste storage and treatment facility was rezoned from industrial to residential at the behest of a politically well-connected developer. Unbeknownst to the owner of the RCRA facility, the developer told the buyers of the $300,000 homes that the facility and other industrial neighbors were going to be moving out of the area. When this didn't happen, the new residential neighbors didn't blame the developer—they blamed and actively sought to oust the RCRA facility and the other industrial neighbors. It didn't matter who was there first or who was at fault for building residences near a hazardous waste facility.

In the case of this particular facility, the entire community relations effort was focused on attempting to coexist peacefully with new neighbors who didn't want them there.

Other issues that have no direct tie to a specific facility, site, or project, but that can create problems that need to be considered in the community relations effort, can include concerns regarding problems or incidents in conjunction with similar facilities, sites, or projects (or those that community members perceive as being similar). This type of adverse publicity needs to be addressed in a community relations plan, since it can lead to erroneous public perceptions that will need to be corrected.

OTHER CONSIDERATIONS FOR DESIGNING A COMMUNITY RELATIONS PROGRAM

In addition to identifying stakeholders and information needs, persons who wish to prepare plans for effective community relations programs must take a number of other factors into consideration. These factors, which appear in the following paragraphs, include:

- How well the program will support statutory requirements for disclosure of information or for public involvement.
- The role played by the history or public image of a facility, site, or project.

- The need for sustainability of the community relations program.
- Timetables to maximize the effectiveness of the community relations program.

Community Relations Programs in Support of Statutory Requirements

Many community relations programs are initiated to support statutory requirements for disclosure of information (reporting of data for inclusion in the Toxics Release Inventory or the reporting of risk management plans under Section 112(r) of the Clean Air Act) or public involvement in conjunction with cleanups of contaminated sites (under CERCLA) or permitting (under RCRA). If a plan is being developed to support these requirements—or such statutory requirements will be major parts of other, all-inclusive community relations plans—then the community relations program should be developed so that it will provide a logical platform of support for the statutory requirements.

A community relations program in support of the permitting of a hazardous waste treatment, storage, and disposal facility under RCRA should include all of the requirements included in the RCRA public involvement regulations. Additional communication and activities that may be advisable to address the needs or interests of the specific community would also be included in the plan; the regulations themselves are fairly comprehensive, however, and form a good basis for a community relations program.

Conversely, community relations programs designed to support risk management plan and "worst-case scenario" reporting will require additional thought on the part of community relations specialists. The regulations promulgated under Section 112(r) of the Clean Air Act call for the reporting of information, which will then be made publicly available; however, these regulations don't prescribe actions facilities should take to talk to their communities about this information. Thus, the community relations plan will have to be prepared out of whole cloth, and community relations specialists will have to decide, based on the characteristics of the facility community (as ascertained through the assessment) and specific information about the facility, the sort of community relations program that will be needed to support the risk management plan disclosure.

Role of History and Public Image

Whether a community relations program is designed to be proactive or as a response to a specific issue, the history and public image of the facility, site, or project will influence the way the program is received. Thus, the community relations program must take these factors into account. For example, if a facility has a bad reputation in regard to environmental performance, the community relations program and the communication that supports it will have to address and correct that reputation. Similarly, if people hold certain beliefs about industry or government in general that have an effect on how they will receive information about the environmental issue in question, then the person designing the community relations program will have to consider these beliefs, as well.

In one case, a community assessment in support of RCRA permitting efforts at a large manufacturing facility revealed that virtually everyone interviewed (including community business people *and* some local government officials) believed that environmental regulations were meaningless, since, in their opinions, big companies do as they please and simply pay fines if they are caught, as this course of action was cheaper than complying with regulations.

This belief created a major obstacle for communicating about the issue at hand—the permitting of an on-site hazardous waste incinerator—since explaining the safeguards built into the regulations governing such units would do little to convince stakeholders that the unit would operate safely. Other methods designed to build personal trust with facility management and employees and demonstrate accountability had to be emphasized in the community relations program.

Organizations that are perceived as untrustworthy face tough community relations challenges. Not only do they have to address stakeholder issues about the situation in question, but they must also confront how the images of their organizations affect stakeholder perceptions and their own credibility. The successful outcome of a community relations program involving such organizations typically hinges on either demonstrating, through some impartial means, that its claims or actions are what it says they are or on finding a third party that is respected by stakeholders and relying on that entity to confirm the validity of the information provided.

A company with serious credibility and image problems stemming from past violations and fines managed to turn its reputation around through an intensive community relations campaign. The linchpin of the effort in rehabilitating its reputation involved volunteering to participate in a pilot program with the state environmental agency to improve both its compliance with the regulations and its overall environmental performance. Management also met repeatedly with community officials, representatives of civic groups, neighboring businesses, and representatives of the local environmental organization to discuss the company's efforts to improve its environmental performance and to obtain their input.

The combination of improved environmental performance and effective communication and relationship building managed to repair the company's poor reputation in the community. The icing on the cake occurred when a senior-level official of the state environmental agency, who, several years earlier, had announced some stiff fines for the company, spoke on the company's behalf at an open house it held to unveil a facility expansion.

Companies or government units that have good reputations and the trust of the public should be mindful of the importance of using good community relations processes. Pre-existing relationships and pre-existing trust and credibility make it easier to handle community questions or concerns in a positive manner—that's the reason we advocate a proactive approach. Trust can be fragile, however, and in situations involving concerns about health or environmental effects, failure to work productively with the community and listen to stakeholders' concerns can chip away at even the best of reputations.

The Importance of Sustainability

Community relations programs that are developed to support a facility, a long-term project, or a site that will take a number of years to clean up should be designed for sustainability. This means that the community relations effort should not end after initial concerns have been addressed or questions answered, even though the amount of communication and activity may be scaled back considerably. Instead, some effort should continue to ensure that lines of communication remain open. It also means that those responsible for community relations efforts should anticipate future concerns or questions and deal with them on a proactive basis. For example, as the investigation and cleanup of a site progresses from the investigative phase to the choice of a cleanup method, or as the cleanup itself progresses, information should be provided to stakeholders on a regular basis apprising them of what is happening.

Timetables for Community Relations Activities

Timetables are important both to ensure that community relations activities in connection with the statutory requirements mentioned earlier in this chapter occur when they are supposed to, and to ensure that lines of communication don't inadvertently close because project or facility managers become involved in other, pressing operational issues.

Timetables for community relations activities often seem less important when a program is designed to respond to an immediate issue. The thought is that communication and activities should be undertaken as soon as possible to defuse the situation. This is true; however, as we stated in the previous section, sustainability of the community relations effort is very important in situations that are ongoing or of substantial duration. Thus, developing a timetable that will ensure that the dialogue continues after the immediate concerns are laid to rest is extremely important. Indeed, few situations involving serious concerns and long duration are resolved in the first few months of a community relations program. Stakeholders may settle down once the dialogue begins; however, cutting off the communication out of an erroneous belief that a complex and contentious issue is settled while it is still, in fact, ongoing can destroy the trust that was built up.

Timetables are important for community relations programs to remain proactive. Timetables for programs should consider events that may be anticipated, such as receipt of data from an investigation of a suspected contaminated site or submittal of a permit application. They should also consider facility or project managers' schedules.

For example, a community relations program will invariably involve making the acquaintance of a number of potential stakeholders, including elected and staff officials, neighbors, leaders of business and community organizations, and so on. When a program is being implemented proactively, before an issue has emerged in the community, it is easier—not to mention more "natural" from a community relations standpoint—to meet these people a few at a time, over a period of time. Thus, community relations specialists will need to sit down with facility or project managers and agree on a schedule that isn't too burdensome, but that will result in making the acquaintance of these potential stakeholders within a reasonable length of time.

Timetables and the scheduling of community relations activities also need to take into account the timing of other events or issues in a community. For example, scheduling a major event, such as a public meeting, right before local elections can risk politicizing the issue that the meeting is about, or lead managers to invest significant amounts of time in working with officials who are then voted out of office. Conversely, events, such as open houses and tours, may be run in conjunction with community events, such as festivals or fairs, that draw a number of people to the area.

One facility manages to bring hundreds of people through its gates for tours by tying its annual open house to a community festival. As part of the festival, several area businesses offer tours—with regularly scheduled buses that shuttle festival goers from facility to facility and back to the center of town. The annual open house is an important event for both the facility and its host community.

Using Scenarios in Plan Preparation to Ensure a Flexible Program

Along with plotting out timetables and building in activities for sustainability of the community relations dialogue, the planning process should consider a range of activities designed to fit alternate scenarios. The first scenario should lay out a "best case" plan for providing information and establishing linkages for a dialogue and relationships in a manner that minimizes concerns. Additional scenarios would be developed based on possible reactions suggested by information gathered during the community assessment.

For example, the analysis of the information collected during the assessment may suggest that certain stakeholders may not be immediately amenable to joining in a dialogue, or that environmental activist groups may be invited into the community to help certain stakeholder groups push for their agendas. Scenarios that would address these possibilities should be considered. The environmental issue itself may also not be well defined at the time that the assessment is performed. For example, the existence of contamination at a site or facility may be suspected, but the amount of contamination or the distance it may have migrated (and whether neighbors' wells have been affected) may not be known at the time the assessment is conducted. Thus, the community relations plan should include a scenario that would provide strategies for working with stakeholders if the contamination hasn't yet migrated off-site—and a scenario in which neighboring wells have been affected or are likely to be affected by the contamination. As we've discussed, depending on the community, these two scenarios—on-site contamination and off-site contamination—may involve very different degrees of concern and the inclusion of different stakeholder groups.

By identifying alternate courses of action based on likely community responses to an issue, community relations specialists won't be too locked into one course of action that may not address the current state of affairs regarding the public's responses to an environmental issue.

Previous Community Relations Efforts and Established Relationships

Planning for a community relations program should also take into consideration relationships that have already been established and actions that facility or project managers have already taken in the way of community relations. In the case of facilities, at least, there are usually some relationships or some past activities that can and should be built upon in a formal community relations program. Facility and project managers do not always recognize the significance of some of the actions they have taken—such as getting to know the local fire chief. Any positive relationships that have been established should be built upon in the formal program. If relationships or experiences have been negative, then these should also be considered. If the negative experiences were with an individual or a group that would be considered a primary stakeholder, then consideration needs to be given as to how the relationship can be mended.

Relationships with stakeholder groups that employees or other people connected with a facility, site, or project have established should not be overlooked, either. Employees who are active in community organizations or who live in potentially affected neighborhoods can be important conduits of information and can help establish sound, personal relationships with formal and informal leaders.

SETTING GOALS AND OBJECTIVES FOR A COMMUNITY RELATIONS PROGRAM

Community relations programs that have clearly articulated goals and objectives are more likely to have focused communication and activities and to be "successful" as measured by both the organization and the community. The goals and objectives of the community relations program need to reflect the goals and objectives of the organization that is engaging in the community relations activities.

It is also important to look at the goals and objectives of stakeholders so that linkages and commonalities among the parties that will be engaging in the dialogue can be identified. These commonalities can be used to identify opportunities to initiate a dialogue on a positive note. In many cases, some commonality of purpose can be identified that can be used to get people to the table. Additionally, identifying those goals or objectives where stakeholders and management of a facility, site, or project are opposed can help community relations specialists understand the types of information they may need to develop or the accommodations they may need to make to address these differences.

The Relationships among Goals and Objectives at These Levels

In this book we talk about goals and objectives at three levels: the organization, the community relations program, and the communication effort or specific communication. The organization itself has goals and objectives. In a private sector organization, these may address profits, market share, the industry in which a company wishes to compete, and the like. The community relations program also has goals—typically, to ensure that public concerns about an environmental issue are addressed in a manner that will result in a positive resolution for the organization.

If the goals and objectives of the community relations program conflict with those of the organization as a whole, then the program cannot "ring true." This is an interesting issue, since many organizations are only just beginning to struggle with the concept of considering the community as a serious force that needs to be accounted for when making operating decisions. It is likely that the needs of community relations programs may, in fact, cause some organizations to revisit and restate some of their current organizational goals and objectives—actions that some public and private sector organizations have already taken.

The goals and objectives of the communication effort or a specific communication should closely follow the goals and objectives of the community relations program. These goals and objectives meld the issues that have been raised (or are likely to be raised) with the needs of the community relations program, the needs of the organization, and the needs of the stakeholders.

For example, owners of a contaminated industrial site plan to clean up the property and redevelop it, also for industrial use.[2] They have the following goals and objectives for this project, for their business, and for the community relations effort, in conjunction with the cleanup:

2. This hypothetical case considers issues likely to arise in regard to risk-based cleanups and redevelopment of property.

XYZ Corp.

Business Goals:

- To clean up the site to minimize the risk of legal action and to allow the property to be available for redevelopment and reuse.
- To minimize the amount of money that must be spent (or maximize financial gain from redevelopment and reuse), while recognizing that agency environmental standards must be met.

Objectives:

- To work with the state agency to prepare workplans and investigate the site, characterize its contamination, design a remediation strategy, and remediate the site.

Community Relations Goal:

- To minimize the likelihood that community residents will file a "stigma suit," or create an environment that is hostile to the ultimate redevelopment and reuse of the property.

Objectives:

- To identify stakeholders and their concerns.
- To promote a dialogue with stakeholders and help them feel comfortable with activities at the site.
- To promote stakeholder support for the remediation strategy and instill confidence in the remediation process in order to maximize acceptance of final test results showing that cleanup objectives have been achieved.
- To promote stakeholder acceptance of plans for redevelopment and reuse of the property by providing information on environmental safeguards that will be put into place to minimize risk of future contamination.

In regard to this issue, a community assessment identified six stakeholder groups. Group 1 is composed of most of the elected and staff officials in the community. Group 2 is composed of many of the business interests in the community, including the local chamber of commerce. Group 3 is a local environmental organization that is particularly concerned about contamination caused by industrial facilities (one elected official and several staff officials are allied with this group). Group 4 is another local environmental organization that is interested in protecting wildlife habitats and streams. Group 5 is composed of residential neighbors, most of whom live in a new development near the site and moved to the area in the past four years. Group 6 is also composed of residential neighbors (including one elected official and one staff official), most of whom live in an area of older homes near the site.

The assessment revealed that the primary goals and objectives of these groups are as follows:

Goals:

- To ensure that residents' health and property values are not adversely affected by either the cleanup—or the levels of cleanup objectives set by the environmental agency—or redevelopment and reuse of the property. (Stakeholder groups 1, 2, 3, 4, 5, 6)
- To maximize the benefit to the community from the opportunity to reuse the property after the cleanup has been completed. (Stakeholder groups 1, 2, 3, 4, 5, 6)
- To maximize the benefit to the community from the industrial redevelopment of the property. (Stakeholder groups 1, 2, 6)

Objectives:

- To push the environmental agency to set very strict cleanup objectives (Stakeholder groups 1, 3, 4, 5, 6)
- To consider placing legal pressure on the owner of the site to ensure that a thorough cleanup is undertaken. (Stakeholder groups 3, 5)
- To consider demanding third-party testing and data verification to ensure that the cleanup is complete and that the site does not pose a risk to residents. (Stakeholder groups 3, 5, 6)
- To ensure that no other industrial facility is sited on the property (Stakeholder groups 3, 4, 5)
- To increase revenues to the community through increased tax revenues from a new facility (Stakeholder groups 1, 2)
- To increase revenues to the community through the creation of new jobs at the new facility (Stakeholder groups 1, 2, 6)

In this case, it is clear that some important commonalities exist: both the owners of the site and all of the stakeholders want the site cleaned up. Differences in the stakeholders' objectives reflect their different agendas, the level of trust they have in the site's owners, and/or lack of knowledge about or confidence in the owner's ability to clean up the site. There are also differences of opinion among stakeholders regarding whether the site's owners should be sued, whether a third party, in addition to the state's environmental agency, should look at the test data (or even take its own samples), and whether or not redevelopment of the site property is desirable.

As should be clear from this example, listing the goals and objectives of all parties involved provides some solid clues regarding the types of information that will need to be made available to stakeholders, as well as the level of effort that is likely to be necessary to achieve a successful conclusion for the site's owners that will be acceptable to stakeholders. We will continue discussion of this case in chapter 6.

EVALUATING THE EFFECTIVENESS OF A COMMUNITY RELATIONS PROGRAM

Organizations that intend to commit to a long-term community relations program should consider measuring the results of the program and its communication effort. The performance of a community relations program should be measured with its goals and objectives in mind. In cases involving permitting or other actions that require some sort of permission from the community to carry out an activity, success may be measured by whether permission—the permit—is granted. In many other cases, however, success involves achieving less tangible goals, such as changing stakeholders' beliefs and attitudes regarding an issue.

One reason to track the progress of a community relations program is to uncover and resolve problems before they become so large that they threaten the goals or objectives of the entire community relations effort. As the following case illustrates, evaluating the success of a program can point up glaring faults.

One strategy in a program designed to increase the number of households checked for radon was to educate children in the classroom about radon's hazards. The children were to be provided with materials to take home to their parents that encouraged home radon testing. The program provided teacher training and classroom materials, but after allowing sufficient time for teachers to complete their instruction, there was no significant increase in requests for home radon tests. The program managers concluded that using children to influence their parents was not an effective strategy to achieve their objective. However, a more careful review of the situation uncovered that the teachers had actually not sent the desired materials home with the children. The teachers had access only to a mimeograph machine, but the materials they were given were suitable only for photocopying, not mimeographing. (From *Communicating Environmental Risk: A Guide to Practical Evaluations*, U.S. EPA. EPA-230-01-91-001. Prepared by Michael J. Regan and William Desvousges, December 1990.)

Determining What to Measure

A number of "results" of a community relations program can be measured. The simplest measures are quantitative, such as the number of calls to a hotline set up to respond to inquiries about an issue. Examples of some common quantitative measures include:

- Number of people who attend an open house or take tours.
- Amount of written material taken from a meeting or exhibit.
- Number of media exposures.

This kind of tracking can provide only a limited amount of information, however, and one should not attempt to interpret too much from these results alone. After all, just because 100 fact sheets were distributed at a public meeting does not mean that everyone who took one necessarily read it, understood it, or agreed with the information it contained.

As a result, simple quantitative measures should be augmented with data on changes in awareness or attitudes *and* with information that can allow community relations specialists to determine whether increased awareness or changes in attitude are having the desired effect—are making the overall goals of the community relations program more attainable. For example, is increased awareness of the measures taken by a facility to protect the environment actually leading to greater comfort with and acceptance of the facility's operations on the part of community residents (the goal of the communication effort), or do residents still oppose the facility because they believe that its location is inappropriate, given surrounding land uses?

Measures of Effectiveness

As we discussed in chapter 3, there are several tools available for gathering information on awareness, level of knowledge, and attitudes, as well as other information on the types of communication vehicles stakeholders prefer, and so forth. These tools include questionnaires, focus groups and small group meetings, and one-on-one meetings. Measurement is obviously more meaningful if baselines have been set at the beginning of a community relations program or communication effort; given the newness of many organizations' community relations programs, however, managers may need to set baselines mid-program—or collect follow-up information after a community relations effort has been completed.

The actual questions used to measure performance—whether in the form of a questionnaire or for use in a focus group or one-on-one meeting—need to generate specific information that speaks to the facility, site, or project itself. As we mentioned in chapter 3, questionnaire design requires special knowledge. Persons who are charged with measuring stakeholder responses to a community relations program should either work with professionals who have expertise in this area or consult the literature on marketing research or opinion polling techniques.

Increased awareness is usually a necessary precondition for any subsequent behavior or attitude change; thus, increased awareness about information surrounding an issue can be an indicator of the effectiveness of a community relations program. Questionnaires, focus groups, and interviews can all be used to measure awareness. As we mentioned, the most meaningful data will measure change in levels of knowledge or awareness by comparing results either midway through a program or at the completion of a program against a baseline. This requires levels of knowledge to be tested at the beginning of the community relations effort, with subsequent surveys conducted as the community relations effort continues (or at the end of the community relations effort).

Attitudes and opinions can also be gauged through interviews, focus groups, or questionnaires—and preferably through a combination of all three. Questionnaires used to measure attitude changes frequently use Likert scales to measure respondents' degree of agreement or disagreement with a series of statements (e.g., "agree strongly," "agree somewhat," "neither agree nor disagree," "disagree somewhat," "disagree strongly"). For such questionnaires to yield usable data, the statements that are being responded to should evoke strong agreement or disagreement so that definitive answers can be collected (e.g., "XYZ Corporation has improved its environmental performance," "Since the cleanup, the ABC site no longer poses a risk to area groundwater").

Another commonly used scale to measure attitudes is the comparative scale, which provides information on relative desirability or performance by asking respondents to rate the subject of the question from "excellent" to "poor" (e.g., "excellent," "very good," "good," "fair," "poor": "How good was the information provided by XYZ Corp. about the cleanup?") As in the case of questionnaires to measure awareness or knowledge, setting baseline measurements—preferably at the beginning of a community relations program—and then surveying additional residents (either specific stakeholder group members or community residents chosen at random) at intervals throughout a program or communication effort will yield information on how attitudes have changed.

Focus groups or one-on-one interviews can provide further insight into attitude changes—including the reasons why attitudes have (or haven't) changed—vital information for ongoing community relations programs.

COMMUNITY RELATIONS PROGRAM PLANNING WORKSHEETS

The following worksheets provide some guidance to persons who must organize and analyze the information they have gathered about an issue and a community. These worksheets consider issues, stakeholders, goals and objectives, timetables, statutory requirements, and resources.

COMMUNITY RELATIONS PLANS WORKSHEET 1
Issues

Describe the issues that need to be addressed:

_____

_____

_____

_____

What is the source of the questions or concerns?

_____

_____

_____

_____

Describe any statutory requirements for the disclosure of information or for public involvement that affect the facility, site, or project in question:

_____

_____

_____

_____

What other environmental or nonenvironmental issues or concerns in the community may have an effect on the environmental issue in question?

_____

_____

_____

_____

COMMUNITY RELATIONS PLAN WORKSHEET 2
Stakeholders

List the primary and secondary stakeholders identified for this issue:

_____

_____

_____

_____

If other potential stakeholders identified during the community assessment are not included, why weren't they included?

_____

_____

_____

_____

For each primary and secondary stakeholder group, describe their interests in/concerns about the issue and their goals and objectives, to the extent known:

_____

_____

_____

_____

_____

_____

_____

_____

_____

_____

Describe any special characteristics of these groups that need to be taken into consideration in the community relations program (e.g., non-English speaking):

_____

_____

_____

_____

Which of these groups can be handled together? Which need to be handled separately?

_____

_____

_____

_____

For each stakeholder group, list channels of communication and activities that would best suit its needs and the types of information to be conveyed:

_____

_____

_____

_____

_____

_____

_____

_____

COMMUNITY RELATIONS PLAN WORKSHEET 3
Goals and Objectives

Community Relations Goal(s):

_____

_____

_____

_____

Reason for Goal(s):

_____

_____

_____

_____

Objectives:

_____

_____

_____

_____

What need will each objective meet?

_____

_____

_____

_____

Describe the activities/communication vehicles required
to meet each objective:

_____

_____

_____

What effect does the environmental issue have on the goals of the organization?

_____

_____

_____

_____

What are the goals or desired outcomes for the issue on the part of each stakeholder group?

_____

_____

_____

_____

How would these desired outcomes affect the organization?

_____

_____

_____

_____

Which goals are shared (or could coexist) among stakeholders and the organization implementing the community relations program?

_____

_____

_____

_____

_____

Which goals are opposed?

_____

_____

_____

_____

Of those goals that are opposed, what are the stakeholders' goals based on?

_____

_____

_____

COMMUNITY RELATIONS PLAN WORKSHEET 4
Timetables and Resources

Describe the timetable for the community relations program or to meet specific goals or objectives within the program:

_____

_____

_____

_____

List any milestone dates that need to be met or actions, events, or impending decisions that will influence or need the support of the community relations program:

_____

_____

_____

_____

For each activity or communication vehicle planned for the community relations program, list the resources that will be necessary (including written materials, personnel time, displays, space rental, postage or delivery charges, etc.)

_____

_____

_____

_____

CHAPTER

6

IMPLEMENTING AN ENVIRONMENTAL COMMUNITY RELATIONS PROGRAM

In the previous chapter, we discussed the process of designing a community relations program and preparing a written community relations plan. This chapter builds on this process by exploring some additional considerations regarding establishing a dialogue with the community. Three key drivers of public concern and opposition—misinformation or lack of information, differences of opinion, and concern over actual incidents or the discovery of risk—are explored, as is the need to avoid adversarial relationships.

One of the major considerations this chapter examines is the concept of stakeholder empowerment. Providing information to stakeholders and engaging in a dialogue with them invests some degree of empowerment in the community. Accommodations—changes in operations or actions taken in response to stakeholder requests—and direct involvement in the decision-making process grant even greater control to stakeholders. Last, this chapter examines several organizations and initiatives that have embraced the concept of a proactive dialogue with the public.

PAYING ATTENTION TO PROCESS

Throughout this book, we have referred to community relations as a process. This concept bears some discussion within the context of implementing a community relations program. In the previous chapters, we discussed assessing the community and stakeholder needs, the setting and use of goals and objectives, and the types of activities and communication vehicles that can be used to develop and maintain a positive dialogue with stakeholders. Although we have introduced each of these elements individually, the community relations process itself involves give-and-take and shifts in attitudes, opinions, and beliefs over a period of time. The adage of community relations specialists to "communicate early and often" refers to the need for a dynamic dialogue, over time, so that stakeholders can process the information provided by the organization responsible for the community relations effort and so that the managers of the organization can consider stakeholders' wants and needs in the decision-making process.

Successful processes need to be shepherded along the way to make sure they stay on track and that needs for information (both giving and receiving) are being met. As we discussed in the last chapter, community relations programs must be flexible to allow for new issues that may arise as information is generated and for the entry of new stakeholders into the dialogue. This means that someone must be overseeing the progress of the dialogue and paying close attention to how it is proceeding. Community relations specialists, who, because of their experience, can often anticipate the pacing of the dialogue or the likelihood that new issues will arise, often fulfill this function. In cases in which the responsibility for a community relations program falls on another manager in the organization, he or she should carefully monitor the goals and objectives of the program, as well as the timetable that has been set for any statutory or organizational requirements. If new issues arise or frustration or discouragement is voiced by stakeholders, then the community relations plan should be revisited to determine what changes need to be made to keep the community relations dialogue from breaking down.

DRIVERS OF THE NEED FOR A COMMUNITY RELATIONS DIALOGUE

As we discussed in chapter 2, each community is unique and the reactions of its residents and other stakeholders to an environmental issue will vary. In analyzing community reaction to an issue, it is important to identify the root causes of concerns in order to determine how best to respond. The authors have identified three basic drivers of concern regarding environmental issues:

- Misinformation or lack of information or understanding about a situation, substance, or process.
- Differences of opinion or belief as to the way a facility, site, or project should be operated or addressed, the risk posed by a substance or activity, or the way resources should be used.
- Incidents, adverse effects, or threats (or perceived incidents, adverse effects, or threats) to human health or the environment.

Environmental concerns frequently involve a combination of these drivers. For ease of discussion, however, we will analyze them singly.

Issues Arising Out of Misinformation or Lack of Information or Understanding

A significant number of issues that lead to public concerns are precipitated by misinformation or lack of information about a facility, site, or project. As we discussed in chapter 1, when it comes to dealing with the public, questions and concerns need to be addressed, regardless of whether they are based in reality or have technical merit. Community opposition is based on *opinions* and *beliefs*, and the obligation to provide accurate information rests squarely on the managers of the facility, site, or project—community residents are *not* obliged to seek this information out on their own.

Concerns based on misinformation or lack of information can be extremely challenging to address. Both the need to establish the credibility of the people providing the information and the need, in some cases, to convey information of a technical nature can take substantial effort. Thus, saying "they just don't understand" doesn't mean that the situation will automatically be simple to remedy. As discussed in chapter 3, the task of changing stakeholders' attitudes means that they must be willing to accept the information they are given. Thus, it is not enough simply to provide information and walk away; community relations specialists may also have to engage in a substantive dialogue with stakeholders to help them "process" this information and feel comfortable with it. Additionally, "accommodations" may have to be made to reflect stakeholders' suggestions or concerns as they learn about a facility, site, or project. Figure 6.1 depicts some of the barriers to correcting misinformation.

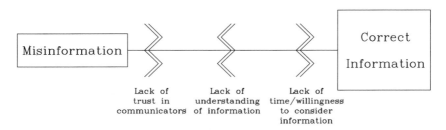

Common barriers to
understanding, accepting
information

Misinformation

Correct
Information

Lack of
trust in
communicators

Lack of
understanding
of information

Lack of
time/willingness
to consider
information

■ **FIGURE 6.1** **Barriers to Correcting Misinformation.**

As information is provided to stakeholders, concerns based on misinformation or lack of information may shift to differences of opinion or belief regarding what should be happening at a facility, site, or project. This scenario is one that many managers throw up as a reason why they would rather not provide information to stakeholders—that, even if the correct information is provided, stakeholders will continue to oppose the facility, site, or project. The authors would argue that, in cases in which this sort of shift occurs, stakeholders are typically moving from flat, unbudging opposition to a negotiation mode, and their increased understanding can be used as a basis for additional dialogue. Thus, this type of shift often *does* represent positive movement, as shown in Figure 6.2, since stakeholders are engaging in a

Opposition   Acceptance

Opposition based on
misinformation or
lack of information

Acceptance based on
knowledge, input into
decision making

Flat opposition lessening; desire
to negotiate, discuss issues increasing

Progression of dialogue

■ **FIGURE 6.2** **Dialogue and Negotiation Allow Additional Exchange of Information.**

positive dialogue based on real information and a workable solution is no longer below the horizon line.

Misinformation is frequently based on a community's experiences or on information stakeholders have received in conjunction with another facility, site, or project, on popular but erroneous perceptions of risk, or on perceptions that do not reflect current regulations or operating standards. For example, residents of a community whose experience with facilities that manage wastes includes landfills are likely to assume that a recycling facility will also be disposing of wastes on-site unless they are told—repeatedly—that this is not the case. If community attitudes about landfills are negative, a recycling facility may be rejected purely on the basis of attitudes toward disposal— even if no disposal is going to take place at the facility. In this case, the recycling facility's developer will have to provide correct information (often repeatedly) about the facility in order to obtain community permission to permit, build, and operate it.

Misinformation can also be based on community residents' remembrances from the past. For example, a facility that used to be a major source of pollution may continue to be viewed as such for years afterward.

In one case, a facility that used to manage wastes and other materials in unlined lagoons created substantial contamination, most of which was cleaned up by a subsequent owner (a system to remove contaminants from the groundwater was installed to clean up the remaining contamination). To comply with regulations to protect the environment, the new owner upgraded all waste and chemical management systems, installing secondary containment and paving over all operating portions of the facility to keep spills that might occur from reaching the soil and groundwater.

Despite the cleanup and the upgrades, community residents still viewed the facility as "dirty," and opposed plans for its expansion. Community residents lacked knowledge both of current regulations designed to protect the environment and the upgrades that had taken place at the facility, and, instead based their beliefs—and their opposition—on how the facility used to operate under the former owners prior to the adoption of many of the current environmental regulations.

Lack of information or understanding is frequently an issue in regard to the investigation and cleanup of contaminated sites, as well as with facilities. Stakeholder concerns about contaminated sites are almost always founded, at least in part, in lack of knowledge about, for example, how groundwater moves through the ground or about routes of exposure. Neighbors hear that a nearby site contains contamination and invariably have questions and concerns about the risks they face. In some cases, simple assurances that risks are minimal given by a credible spokesperson can allay residents' fears. More often, however, additional information may need to be communicated to educate stakeholders about topics such as routes of exposure and the investigation and cleanup process to provide a level of confidence that the actions being taken to protect them and to fix the problem are appropriate and to enable stakeholders to make meaningful comments and suggestions.

Issues that Involve Differences of Opinion or Belief

We have just discussed situations that involve actual erroneous opinions or beliefs based on misinformation or lack of information or understanding. Many environmental issues that require a dialogue to address or resolve community concerns or opposition also involve stakeholders who have a reasonable (or excellent) understanding of a situation, but who believe that it should be handled in a different manner (see Figure 6.3). Such beliefs may include the following:

- A substance or activity poses greater risk to human health and the environment than scientific studies currently suggest.
- Certain hazardous chemicals should not be stored or used at a particular facility—or they should not be used at all.
- Wastes should be recycled or reduced as much as possible, and, to encourage this, additional treatment and disposal facilities should not be built.
- Certain land should not be developed or used for the harvesting or extraction of natural resources.

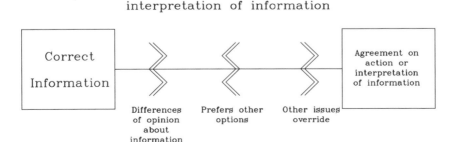

■ **FIGURE 6.3 Barriers to Agreeing on an Outcome.**

■ The placement of a facility is inappropriate or certain activities or operations related to it are inappropriate.

The last bullet point is basically the "NIMBY"—Not In My Back Yard—response. Although a component of NIMBY may be misinformation or lack of information, the essence of the response is the belief that the facility or project in question shouldn't be put in a particular location that impinges on the concerned parties' personal interests. These people frequently accept the premise that a facility or project is needed—indeed, they may push for a roadway or a landfill, because they perceive the need for it—however, they believe that it should be "elsewhere."

When dealing with differences of opinion, it is important to know what the opinions are based on. For example, are they based on ideology or on concern about property values? Persons who are termed "activists," as opposed to "involved citizens," are frequently pursuing ideology as much as specific results in regard to a specific situation. For example, some activists may oppose incineration or the use of certain chemicals, such as those containing chlorine, *everywhere*—not just in a specific community. Knowing that you will be starting from an absolute point (e.g., no incineration, no chlorine) when working with them will help the community relations specialist craft an approach that minimizes adversarial, head-on collisions with ideology by keeping the dialogue focused on the specific situation rather than on tackling core beliefs. Additionally, establishing a dialogue regarding less volatile issues, or tackling problems where some common ground already exists, can provide the forward momentum necessary to tackle the tougher, ideologically burdened issues with some reasonable likelihood of a positive outcome.

In one case, certain chemicals handled by a facility were of serious concern to some residents of the community. Several of these residents had called for a ban on their use, conducting research to determine if it was possible to pass a local ordinance prohibiting the presence of these chemicals within a given distance of residences. The facility could not discontinue the use of the chemicals in question without shutting down completely. Facility management viewed the residents as extremists and, initially, refused to talk with them, leading to escalation of concern and anger on the part of residents. Eventually, facility managers determined that they needed to begin a dialogue with these neighbors or face serious opposition when a current RCRA permit came due for renewal.

A community assessment revealed that residents were unwilling to discuss the facility's need to use the chemicals or to listen to any information regarding safety precautions in handling them. Thus, there seemed to be little chance of even opening a dialogue. Residents *were* anxious to discuss emergency response capabilities with the municipality, the local emergency planning committee, and facility management, however. The residents wanted to know what to do in case of an emergency involving other chemicals at the facility or emergencies involving a nearby railyard. Opening a dialogue on emergency response addressed a very real issue—and got the parties talking and working together.

If the belief concerns factors such as the effect of a facility, site, or project on property values or other "quality of life issues," the community

---

Ideology and the Community Relations Dialogue

We will digress for a moment and discuss ideology and "activists." Public and private sector managers often criticize activists for their absolute and inflexible beliefs. It is worth considering that all of us hold strong beliefs—and those who work in industry (as the authors of this book do) can also respond in a decidedly visceral fashion when they believe that their ability to compete freely in the marketplace, or to operate in a manner that allows maximum productivity, may be curtailed. Public and private sector managers become very concerned about attempts by the public to curtail their operations both because these attempts may have an impact on them in regard to a specific facility, site, or project, *and* because it rubs up against *their own* ideological constructs about the free-market system.

The courtly dance of collective versus individual rights is played out in the federal, state, and local legislative processes every day. It is also at the core of the controversy over many environmental issues. In the community relations process, ideologies and beliefs on all sides need to be acknowledged, and then set off to the side as much as possible while examining the specific environmental issue in question. Does one abandon one's core beliefs—or the need to clean up a site, protect the environment, or operate a business? No. But the participants in the dialogue should attempt to focus on the issues specific to the facility, site, or project and not hide blindly behind an absolute belief, whether it's "no incineration, ever," or "no interference in private industry, ever."

relations effort may include making some sort of accommodation. For example, a berm or fence may be built, or trees and shrubs planted between a facility, landfill, or quarry and a residential area to cut down on noise and improve the view for the neighbors.

In some cases, differences of opinion or belief may be modified through the provision of additional information on the subject at hand. For example, providing information on the safeguards built into chemical handling procedures at a facility may give persons who are against the use of certain chemicals enough confidence in the way the chemicals will be managed at a particular facility to decide not to continue their opposition. Notice that the community relations effort did not seek to change the stakeholders' core beliefs, just their belief about a particular facility.

In one case, an activist who was well known for his opposition to the use of certain chemicals was recruited to sit on the community advisory panel (CAP) of a facility that manufactured chemicals. His questions were some of the toughest put to the facility's management. Although he continued (and continues) to actively work against the use of certain types of chemicals (including some of those used at the CAP's facility), and worries about the way the *rest* of industry manages chemicals, based on the information he received by participating on the CAP, he believes that the facility in question is doing a good job at managing its operations, and that this facility, at least, does not pose undue risk to its residential neighbors.

There are times when opinions are based on other overriding issues. For example, residents may oppose the construction of a facility because it would require a zoning change that would curtail the use of their property for some purpose that they desire. Such situations, which pit the concrete self-interests of various parties against each other, may not be resolved through dialogue only. These types of scenarios are common in the construction of roadways and other public works projects, and can lead to serious and protracted opposition that may have to be resolved through the powers of government or through the court system.

Issues that Involve Incidents, or Damage or Threat to Human Health and the Environment

Concerns about environmental issues may also be triggered by actual incidents, such as spills, or the discovery of adverse health effects from certain substances, such as asbestos, either at the facility, site, or project in question or elsewhere. The publicity over discovery of contamination in residential areas, such as Love Canal, New York and Times Beach, Missouri, has sensitized people in many communities throughout the country to potential environmental hazards.

Community relations specialists need to address accidents, adverse effects, and the like straight on if they are going to get past them. When you know that stakeholders are concerned about a past incident, it is best to bring up the incident at the beginning of a meeting or conversation, deal with it, and go on. There are several reasons for this. First, bringing up the incident yourself disarms critics and provides greater control about how the incident is discussed. Second, if an incident is not mentioned, but is prominent in people's thoughts, it serves to distract people from discussing other important issues. Bringing up the incident yourself can provide at least temporary closure so that it is possible either to discuss it in a nondefensive mode or to move on to other topics.

Depending on the seriousness of the accident or damage, and, often, its causes (gross negligence is a definite sticking point), concerns stemming from incidents can actually be easier to handle *in terms of the actual communication process* than issues involving misinformation or differences of opinion, because incidents involve concrete, rather than what are often hypothetical, issues. An incident or adverse impact is a "thing" that draws active attention and can lead to an active dialogue. If this dialogue is handled well by community relations specialists, *and* appropriate measures have been taken to address or correct the incident or adverse impact, then concerns regarding the specific incident itself can often be addressed and may be resolved. However, this dialogue may reveal other concerns (such as differences of opinion on chemical use) that must then be dealt with. Thus, dialogues about specific accidents may evolve to address other concerns.

As we discussed in chapter 3, community relations in the wake of a major accident or discovery of risk will involve communicating with stakeholders about that risk and helping them understand the impact it has had or could pose to them. A good understanding of risk communication techniques is important to ensure that this information is conveyed in a manner that is accurate and provides the necessary information, but that doesn't create undue alarm or fear. This information may need to be repeated several times, and in different formats, to allow stakeholders to process it. Additionally, it is likely that stakeholders will want additional information to help them understand what happened and why, as well as what is being done to remedy the situation, if they are to feel that they (or the environment) are being protected from undue risk. For example, in cases involving discovery of contamination or suspicions that contaminants are present at a site, nearby residents or others who come near the site will invariably want to know that they are safe—and they may need to be reassured as to their safety throughout an investigation or cleanup.

If some stakeholders have been personally affected by an incident, their concerns and their emotions will often be very strong. Strong emotions can also be present in persons who were not personally affected. These people may be empathizing with those who were affected (or with wildlife or natural areas that have been harmed or damaged). They may feel a threat to their own security or the security of others whom they care about as a result of the incident. They may also feel anger at the organization that caused the damage or injury, perceiving that it lacks concern, or that profits or other goals are more important than the "by-product" of damage to human health and the environment. The anger and concern of these people should be treated with empathy and understanding.

As we discussed in the section on crisis communication in chapter 3, disclosing information and taking responsibility are extremely important to the successful handling of community concerns regarding an incident. Residents and the media often pay as much attention to the way an incident or news of a problem is handled as they do to the incident itself. Ignoring or minimizing an incident or damage (or perceived damage) isn't going to make it go away—and may actually serve to perpetuate anger or concern that would have dissipated if proper closure had been given to the issue.

For example, during a community assessment conducted by one of the authors, residents living near a facility that had experienced a minor fire indicated that they weren't angry about the fire, since "accidents do happen," and no one was injured, but they did look askance at the plant manager, who had, several months after the fact, referred to the incident during a speech to a local civic group as "an alleged fire." As the neighbors pointed out, there had been no "alleged" about it—why did the plant manager suddenly feel the need to act as though it didn't happen? To them, the plant manager was backing away from his responsibility for the facility and to the community, an act they found very disturbing.

Combinations of Drivers of Concern

As we noted, concerns surrounding environmental issues often have several drivers. Misinformation or lack of information often drives differences of opinion or belief. Concern over incidents, environmental damage, or risk to human health is often amplified by misinformation or lack of information, and can drive the adoption of beliefs or opinions regarding real or perceived threats and the organizations that are responsible for them.

How should such issues, which involve several drivers of concern, be handled? It is usually best to tackle the problem of misinformation or lack of information first, to lay the groundwork for a productive dialogue. Although some of this information may not initially be accepted, either due to lack of trust or because of differences in opinion or belief, it is important to get correct information out there at the outset. Repetition of messages, the building of substantive, positive relationships, and confirmation of information (or dissemination of information) by respected third parties can often overcome much of the misinformation that can accompany early concerns—at least among most stakeholder groups. In cases marked by strong controversy, emotion, and lack of trust, however, acceptance of correct information can take time.

Once the process of correcting misinformation or providing missing information is under way, attention should be turned to addressing differences of opinion or concerns about actual incidents. In the case of addressing actual incidents, some closure often needs to take place. This will often involve sitting down one-on-one or in small groups and discussing what happened, why and how it happened, and the measures that have been taken to prevent such incidents or damage from happening again. Obtaining stakeholders' input by listening to their concerns is extremely important. And taking action to prevent future incidents is of paramount importance. As we've said, community relations is not a panacea—nor should it be used as a placebo in place of addressing real problems that need correction. In the case of accidents or other damage, the ultimate "proof" to stakeholders will be the subsequent track record. For example, if a spill occurs, the community relations dialogue can provide information that can minimize concerns; however, what happens over time will convince residents that the organization is "walking the talk."

Achieving Nonopposition Resolving differences in belief or opinion calls for an open and nonjudgmental dialogue. If stakeholders continue to hold differences of opinion after most of the less controversial facts have been brought forth, discussed, and agreed to, and the scrutiny of additional, existing information is unlikely to move the parties any closer together, then community relations specialists must typically aim toward moving stakeholders into a position of "nonopposition." In nonopposition, the parties essentially "agree to disagree" and, it is hoped, move forward from there.

While nonopposition is less desirable than acceptance and support, in the case of many environmental issues, nonopposition is the most positive resolution possible. Opinions and beliefs being as diverse as they are, most environmental issues that are successfully "resolved" involve some stakeholders who aren't in complete agreement. These stakeholders may, however, move into positions of nonopposition either because they believe that the issue cannot be successfully challenged because others in the community have accepted it, or that, although the facility, site, or project is not being handled exactly as they would wish—they may wish to close down the facility or project entirely, if they had their preference—it is being managed as well as something "like that" can be managed.

This leads us to another point about persons who aren't—and will never be—supporters. Some stakeholders will continue to oppose operations at a facility, site, or project despite the best community relations efforts. In such cases, it is best to concentrate on building relationships and support among other stakeholders rather than putting an excessive amount of effort into changing the minds of those who are deeply entrenched in their views.

AVOIDING ADVERSARIAL ENCOUNTERS

One of the objectives of a community relations program should be to avoid, defuse, or minimize adversarial encounters between the organization responsible for the program and stakeholders. If possible, it is desirable to avoid, defuse, or minimize adversarial positions *among* stakeholders, as well, although the organization responsible for the community relations program can often only do so indirectly by setting the tone for a positive dialogue and giving everyone a chance to speak and be heard.

We have discussed the fact that concerns over many environmental issues are often based on differences in opinion or belief—differences over which people can definitely dig in their heels. Additionally, the emotional component of many environmental issues can create situations in which people can become extremely upset and angry. Thus, skillful handling of the community relations dialogue requires a complex maneuver: setting some common objectives that can form the basis for discussion while acknowledging and working with people's emotions so that they do not overwhelm the process.

The first rule for minimizing adversarial postures is to avoid treating stakeholders as adversaries. Although this rule may sound self-evident, it can be tough to stick with if stakeholders are particularly strident or appear to be behaving in a patently unreasonable manner. It can also be difficult because managers who are responsible for a facility, site, or project are themselves usually under pressure to meet goals and objectives that have nothing to do with the sorts of concerns that stakeholders may be voicing. Still, behaving in a confrontational or combative manner toward stakeholders will guarantee that they will return the favor, and lock the "dialogue," such as it is, into an unproductive fight.

Possessing knowledge about stakeholders' agendas and their goals and objectives regarding a specific issue can help facility and project managers and community relations specialists identify common goals or objectives that can provide the basis to establish a positive dialogue that can minimize adversarial encounters. Possessing knowledge about residents' lives, their concerns, and their fears can also help managers understand why compassion and empathy are important attributes when dealing with the public about environmental issues.

Establishing Common Goals and Objectives and Working Through Differences

In chapter 5, we looked at the goals and objectives of a business and several stakeholder groups regarding the cleanup and redevelopment of an industrial site. We will revisit these now to explore how common goals and objectives might be identified and used to promote a positive dialogue.

XYZ Corp.

Business Goals:
- To clean up the site to minimize the risk of legal action and to make the property available for redevelopment and reuse.
- To minimize the amount of money that must be spent (or maximize financial gain from redevelopment and reuse), while recognizing that agency environmental standards must be met.

Objectives:
- To work with the state agency to prepare workplans and investigate the site, characterize its contamination, design a remediation strategy, and remediate the site.

Community Relations Goal:

- To minimize, through the use of community relations, the likelihood that community residents will file a "stigma suit," or create an environment that is hostile to the ultimate redevelopment and reuse of the property.

Objectives:

- To identify stakeholders and their concerns.
- To promote a dialogue with stakeholders and help them feel comfortable with activities at the site.
- To promote stakeholder support for the remediation strategy and instill confidence in the remediation process to maximize acceptance of final test results showing that cleanup objectives have been achieved.
- To promote stakeholder acceptance of plans for redevelopment and reuse of the property by providing information on environmental safeguards that will be put into place to minimize risk of future contamination.

In regard to this issue, a community assessment identified six stakeholder groups. Group 1 is composed of most of the elected and staff officials in the community. Group 2 is composed of many of the business interests in the community, including the local chamber of commerce. Group 3 is a local environmental organization that is particularly concerned about contamination caused by industrial facilities (one elected official and several staff officials are allied with this group). Group 4 is another local environmental organization that is interested in protecting wildlife habitats and streams. Group 5 is composed of residential neighbors, most of whom live in a new development near the site and moved to the area in the past four years. Group 6 is also composed of residential neighbors (including one elected official and one staff official), most of whom live in an area of older homes near the site.

The assessment revealed that the primary goals and objectives of these groups are as follows:

Stakeholders' Goals and Objectives

Goals:

- To ensure that residents' health and property values are not adversely affected by either the cleanup—or the levels of cleanup objectives set by the environmental agency—or redevelopment and reuse of the property. (Stakeholder groups 1, 2, 3, 4, 5, 6)

- To maximize the benefit to the community from the opportunity to reuse the property after the cleanup has been completed. (Stakeholder groups 1, 2, 3, 4, 5, 6)
- To maximize the benefit to the community from the industrial redevelopment of the property. (Stockholder groups 1, 2, 6)

Objectives:

- To push the environmental agency to set very strict cleanup objectives. (Stakeholder groups 1, 3, 4, 5, 6)
- To consider placing legal pressure on the owner of the site to ensure that a thorough cleanup is undertaken. (Stakeholder groups 3, 5)
- To consider demanding third-party testing and data verification to ensure that the cleanup is complete and that the site does not pose a risk to residents. (Stakeholder groups 3, 5, 6)
- To ensure that no other industrial facility is sited on the property. (Stakeholder groups 3, 4, 5)
- To increase revenues to the community through increased tax revenues from a new facility on the redeveloped property. (Stakeholder groups 1, 2)
- To increase revenues to the community through the creation of new jobs at the new facility on the redeveloped property. (Stakeholder groups 1, 2, 6)

In this hypothetical case, there is a very strong common goal among the stakeholders and between the stakeholders and XYZ Corp.—to clean up the site. Although all groups are also interested in beneficial reuse of the property after cleanup, a review of the subsequent objectives reveals that there are some differences of opinion as to what this beneficial reuse should be.

An area of possible contention among several of the stakeholder groups—and definitely between stakeholder groups 3, 4, and 5 and XYZ Corp.—concerns the redevelopment or reuse of the site. These stakeholder groups have stated that they do not want the property redeveloped for industrial use after cleanup, while several of the other stakeholder groups are backing industrial redevelopment because they are looking for increased tax revenues and additional jobs for the community.

Additional areas of possible contention include the allowable levels of residual contamination (the levels to which the site is to be cleaned) and the desire by some stakeholders for third-party verification that the cleanup has been completed.

Minimizing Adversarial Encounters Regarding XYZ Corp.'s Cleanup It is clear that all parties want the site cleaned up. Using the common goal of cleaning up the site as the starting point for a dialogue with stakeholders will minimize the adoption of adversarial positions. The challenge in establishing a productive dialogue will be, first, communicating that XYZ Corp. wants the cleanup (never take for granted that people have heard facts such as this, or that they believe it), and, second, keeping the dialogue from being derailed before it is under way by controversy over the other issues involved—notably, the redevelopment of the property.

It is likely that much of the concern over cleanup standards (and over whether a lawsuit is filed or third-party verification is needed) is founded in lack of information and lack of trust in XYZ Corp. The lack-of-information issue can be addressed by providing information on investigative procedures and analytical and cleanup processes. The effect of lack of trust in XYZ Corp. can be minimized by bringing in a respected third party to talk about these procedures. Since, in this case, the state environmental protection agency's staff will be overseeing the work and setting the standards, it is the logical third party to bring in—indeed, it really must be a part of the dialogue. If the state agency is not viewed as credible by many of the stakeholders, however, efforts should be made to bring in an additional authority to amplify and verify the messages.

Since XYZ Corp. wants to use a risk-based approach to clean up the site to minimize costs, extra effort will undoubtedly be necessary to explain the risk analysis process and how residents will be protected from contaminants that are left at the site.

Cleaning up the site to levels of residual contamination that are suited for industrial redevelopment is a double-edged sword. On the one hand, stakeholder groups 1, 2, and 6 want the property redeveloped for the economic benefit that it will provide to the community. In their cases, the question is whether they will view a cleanup that may leave some contamination in the ground as acceptable. Note that stakeholder groups 1 and 6 join with 3, 4, and 5 in wanting strict cleanup objectives. Acceptance of a risk-based approach by groups 1 and 6 will likely depend on whether the information provided to them on risk analysis and on the measures taken to prevent exposure of area residents to the remaining contamination is sufficiently adequate to allow them to feel confident that the risk-based approach is acceptable and consistent with the objective of redeveloping the property for industrial use.

On the other hand, in the case of stakeholder groups 3, 4, and 5, which oppose industrial redevelopment of the site and appear to possess a significant amount of distrust of XYZ (as evidenced by consideration of a lawsuit by groups 3 and 5 and by their desire—along with the otherwise supportive group 6—for third-party verification of cleanup), use of a risk-based approach is likely to be considered completely unacceptable, at least initially. They are likely to use XYZ Corp.'s desire to use a risk-based approach to cleaning up the site to bolster their argument about why very stringent cleanup objectives should be set and a third party should be allowed to verify cleanup. They are also likely to use the fact that the site isn't going to be cleaned up to "pristine" status as another reason why the site shouldn't be redeveloped for industrial use. A very negative turn of events would be for these stakeholders to indeed file a "stigma suit," which could complicate or block the cleanup process.

In regard to opposition to the redevelopment of the site for industrial use, the question that needs to be asked is why groups 3, 4, and 5 oppose this action. Do they have another proposed use for the property? For example, in the case of stakeholder group 4, is the property contiguous to an area that its members wish to preserve for wildlife? Do these stakeholders (notably groups 3 and 5) believe that an industrial use is out of character for the area? Do these stakeholders worry about industrial processes, such as chemical use or "housekeeping" practices? Do they believe that industrial reuse of the property will lead to more contamination?

Some of these issues may be resolved either because they are based on misinformation (e.g., perceptions of "housekeeping" and chemical storage based on practices used prior to current regulations), or through modifications or accommodations to the design of the final project. Although XYZ Corp. will not want impractical constraints placed on the redevelopment of its property, such modifications may make the redevelopment acceptable to the majority of stakeholders and be worth the cost.

As this simplified case suggests, an appropriate dialogue for this situation should begin with the common objective (cleaning up the site), and it should proceed into an information sharing mode, preferably with the assistance of the state environmental agency and other third parties that can confirm and amplify the information and provide some of the credibility that XYZ Corp. lacks among stakeholders. Small group meetings, one-on-one sessions, workshops, availability sessions, and fact sheets could all be used. Equal emphasis should be placed on listening and receiving feedback and on providing information.

Depending on the level of comfort stakeholders achieve through the information sharing activities, XYZ Corp. might consider providing funding for additional testing, verification, or review of the cleanup activities and risk assessment by a third party. The third party could provide an extra measure of comfort to all stakeholders. A report from a third party could move these stakeholder groups into positions of nonopposition regarding the use of a risk-based approach to the cleanup.

Finally, XYZ Corp. would need to listen closely to stakeholders who are opposed to the redevelopment of the property to understand why they oppose it. If misinformation or lack of information is the problem, then the same measures (information sharing, involvement of a neutral third party) used to address concerns about the cleanup can be used to address concerns about redevelopment. If the problem is differences in belief or opinion regarding the desirability of industrial redevelopment, then efforts should be made to determine what aspects of the redevelopment are at issue. Although it may not be possible to accommodate stakeholders' desires (e.g., a 2,500-foot buffer from the nearest house), insight into their concerns may lead to design changes—such as a naturally planted buffer zone near an adjacent creek—that could make the project more acceptable to them. Some stakeholders who were against the redevelopment may then move into positions of nonopposition, providing XYZ Corp. with sufficient support to comfortably move forward with their project.

Community Activists Versus "Average" Community Residents

A discussion on avoiding adversarial relationships must necessarily address the topic of working with activists. This section will examine community (as opposed to national) activism. At several points throughout this book we have talked about "activists" versus "average" community residents. Earlier in this chapter, we stated that persons usually described as activists are typically pursuing an agenda as well as an outcome involving a specific situation. In the case of "national" activists, this is certainly true. In the case of community activists—who may take on a variety of projects, not just environmental issues—this is also often true. Being "active" in the community, acting as a "watchdog," or pushing for the rights of residents (or certain segments of the community) is a way of life for many of these people. They are often quoted in the newspaper and have, if not a following, then at least the recognition of other residents as persons who put themselves forward. These individuals may be respected or viewed as crackpots, depending on their own personalities and the viewpoints and agendas of those who are asked about them, but they are definitely forces to be reckoned with within the community relations process.

Another type of community activist that community relations specialists often encounter in conjunction with serious issues is persons who haven't had a history of activism per se (they weren't "known" community opinion leaders), but who have become dominant stakeholders with an activist bent because an issue has had a direct impact on them. For example, people who have experienced adverse health effects (or who have family members who have experienced adverse health effects) that they believe are the results of environmental degradation from a facility or site may assume an activist role. As a result of their experiences, some of these people may begin to address other issues in the activist mode, as well.

Other "average" (nonactivist) community residents don't become activists as we have used the term in this section, although they may be vigorously opposed to or concerned about the situation at a facility, site, or project and, temporarily, at least, adopt an activist stance. These people become concerned because they perceive that they are personally affected by an issue. As we discussed in chapter 1, it *is* possible to "awaken the sleeping giant"—to mobilize a large percentage of the affected or potentially affected "average" residents in a community—if residents believe that a situation poses a danger to them, their families, their livelihoods, or their property.

Activism and the Structure of the Community Decision-Making Process

As we discussed in chapter 2, the way community residents respond to threats or perceived threats has a lot to do with the social structure and the mode of decision making in the community. Although most communities have some activists—indeed, activists on various issues often provide the impetus for important community programs or improvements—the way activists fit into the power structure of the community will depend on the community's decision-making dynamics. Communities that allow limited access to the decision-making process, or in which certain groups are essentially disenfranchised, are likely to have many activists or very belligerent activists, simply because activism is their only way of making an impact within an unresponsive power structure. Such activists are prone to grandstanding and to using issues to pursue their own agendas because this is how they "make it up on the radar screen" of the decision makers. Indeed, part of their agenda is often to use issues to gain power within the decision-making structure. Thus, dealing with community activists who are functioning in these kinds of settings can be extremely difficult, since the environmental issue is often only a minor part of their agenda—using the issue to further the rest of their agenda is the other part.

Other, more inclusive communities typically have fewer "activists" but more "involved citizens," because such involvement is valued and positively reinforced. Is there much difference between "activists" and "involved citizens"? In terms of level of interest, no. Ideologically speaking, possibly—at least in terms of how vehemently individuals cleave to certain views. In terms of ease in establishing a positive working relationship, the difference can be very significant. People who do not feel that they are the "underdogs," which the "involved citizens" definitely do not, are going to be coming to the table with a lot less baggage. And even if they ask all of the same questions as "activists," these people are typically viewed differently by facility or project managers and treated a lot better. Stepping back and looking at the context in which "activists" and "involved citizens" function can provide important insight into the challenges or strengths they have to deal from, and can help community relations specialists cut through the rhetoric to ascertain these people's true interests and agendas, as well as their roles in the community. This objective view lessens the likelihood of adversarial encounters and paves the way for a positive and productive dialogue.

As we discussed in both chapters 2 and 5, some communities also seem more amenable than others to inviting in or tolerating "outside" activists. In addition to the issue of attitudes toward outsiders, the way outside activists will be treated—and how residents will expect them to act—often tracks fairly closely with the way local activists are treated or act within the established social, power, and decision-making structures of the community.

Facility, site, and project managers sometimes talk about who has more "validity" in the community relations dialogue—activists, "average" residents, "involved citizens" (i.e., opinion leaders), or formal leaders, such as elected officials. Managers in organizations experiencing community concerns about environmental issues often look upon certain stakeholder groups as having greater rights in regard to participating in the dialogue. While it is true that the opinions and desires of neighbors or other persons who are directly affected by an issue and local elected and staff officials who are responsible for overseeing the public welfare should carry substantial weight in the community relations process (as primary stakeholders), dealing with the issue within the context of the *community* means that others who are interested, such as "activists" or "involved citizens," ought not to be blocked out of the process.

Within the social dynamic of communities, these people are fulfilling their roles—which is why we suggest in chapter 5 that they be afforded primary stakeholder status.

Although involvement of activists or involved citizens may add complexity to the community relations process, attempting to exclude them, particularly from serious issues, can create new problems, since many residents look to these people to act as watchdogs or provide representation to the community. Thus, attempts to exclude them could lead to additional activism or adverse public opinion directly against the facility, site, or project for trammeling on the unwritten roles of these stakeholders.

THE ISSUE OF EMPOWERMENT

A major component in many situations involving concerns about the environment is empowerment and control. People often perceive more risk in actions or situations over which they have limited control. As we discussed in the section on risk communication in chapter 3, people may be willing to take on risk that is of their own choosing (e.g., mountain climbing, skiing, eating a high-fat diet), but they have a very different reaction to risk that is imposed on them. It is not unusual for community relations specialists to encounter chain smokers who express substantial anger over the presence of chemicals well below drinking-water standards in their tap water. Part of the problem is that these people aren't really sure of what the presence of the chemicals means or the amount of risk they actually pose, but the other part of the equation is that they have no control over the presence of the chemicals in their tap water, while they do have control over whether they continue to smoke cigarettes.

Many of the objections to the methods used to clean up sites or operate facilities stem from these same issues—lack of understanding (which we have discussed at length) and concern over lack of personal control. For example, how do stakeholders *know* that the persons responsible for a facility, site, or project will make the right decisions? What if employees make a wrong decision (e.g., open the wrong valve, fail to properly regulate the emissions from the mobile incinerator, cut down the wrong lot of trees)?

Thus, many stakeholders seek to exert their own control over the situation by simply prohibiting certain activities that they perceive as too risky for their comfort. Common statements that stakeholders make to explain why outright prohibition is the only answer include comments such as: "They can't be 100 percent sure," or "There's no way they can guarantee that it won't

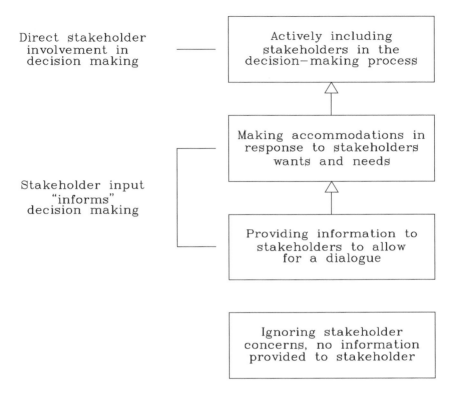

■ **FIGURE 6.4** Hierarchy of Stakeholder Empowerment.

happen." These statements are typically founded, at least in part, on lack of personal control.

Hierarchy of Stakeholder Empowerment

Providing stakeholders with additional information is the first step in helping them gain a level of comfort regarding the issue at hand (see Figure 6.4). In cases involving facility operation, for example, it may not be possible or desirable to make the kinds of changes that some stakeholders would want. Stakeholders may want a facility to cease using certain chemicals that are necessary to the production of goods (e.g., one cannot make polystyrene without styrene, or many pharmaceuticals without chlorine). In these types of situations, providing information about safeguards for the management of the toxic chemicals and information on why the chemicals are needed in the process may be all that a facility's management can do. In many cases, good explanations coupled with good track records in regard to environmental performance and ongoing "maintenance" communication is sufficient to satisfy most, if not all, stakeholders, or at least to move them into positions of nonopposition.

Making accommodations that are important to stakeholders is the next step in the hierarchy of empowering stakeholders. As we will explain later in this section, accommodations are changes in operations or other actions that meet stakeholders' needs in regard to concerns or convenience.

The final step in the hierarchy is direct stakeholder involvement in the decision-making process under or beyond current statutory requirements. The degree to which the decision-making process can or will be made open to public input will, of course, vary depending on the situation. Most private sector operations are only nominally open to public input in conjunction with certain statutes (e.g., local zoning ordinances, requirements for permitting). The "business" side of private sector operations is even less open to public input. Most public sector organizations are more open to stakeholder input because of procedural and statutory requirements (e.g., referendums, open meeting requirements) and must allow for considerably more public input into the decision-making process regarding the expenditures of tax dollars.

If certain stakeholders cannot actually close down or seriously harm operations, then most organizations will probably opt to avoid including them directly in actual decision making and, instead, choose to listen to questions and concerns and provide information (engage in a dialogue), and, possibly, make some accommodations. The organizations may also consider stakeholder suggestions or requests during the decision-making process—stakeholder input will "inform" the decision-making process—however, stakeholders would have no binding say in the final decision.[1] In many cases, this scenario can satisfy the majority of the stakeholders involved.

Inviting stakeholders—especially members of the public—to make suggestions to inform the decision-making process is the most frightening aspect of community relations for both public and private sector managers. After all, what if the public suggests or demands something that would be inconvenient or that would make a project nonviable? Will these people become even more angry if organizations don't do as they suggest than if managers just don't ask them for their opinions at all?

---

1. In the broad context of environmental policy, it can be argued that private citizens do have input into a wide variety of private industry's activities through the legislative process. Statutes that protect human health and the environment by regulating corporate behavior essentially reflect the will of the citizens.

The truth is, if an issue is one of great importance or concern to the community, suggestions or demands are going to be offered whether they are asked for or not. Suggestions (or demands) that aren't accompanied by a dialogue are simply more likely to be even further off the mark and lead to greater frustration or activism on the part of community residents. And listening to suggestions or concerns *can* encourage managers to look at their operations in ways that can lead to improvements.

The authors are not suggesting that managers of organizations cede control of their operations to community stakeholders. Ultimately, public and private sector managers have to make decisions based on their expertise in the operation of a facility, site, or project and the goals and objectives of the organization, since accountability and responsibility for the facility, site, or project reside with them. Although stakeholders may be knowledgeable and, as the premise of this book holds, deserve a hearing, the organizations' managers generally know more about the technical aspects of their operations, and typically have (and should have) the final say.

Still, as we have explained, the line between the organization's needs and rights and the rights of the public (and of individuals) is not always clearly drawn—which is the reason why the need for community relations exists. The question of who is in control, and to what extent, is the stuff of many heated exchanges between the representatives of public and private organizations and stakeholders and, in fact, the successful resolution of serious concerns regarding environmental issues often involves granting community residents a measure of control beyond simply the provision of information.

Accommodations

Accommodations are changes that are made or actions that are taken at a facility, site, or project in response to suggestions that stakeholders make to increase their level of comfort or to meet their needs regarding an issue. The goal of an accommodation is to achieve a compromise—a solution that works and that ideally doesn't place a disproportionate burden on any one party. As such, accommodations are an important aspect of community relations. For example, in chapter 3, we cited an excellent example of an accommodation—an air monitor in a residential neighborhood that could be read by anyone who wished to check whether emissions from a nearby mobile incinerator were above regulatory limits set to protect human health. This accommodation added little to the overall project cost and did not affect the cleanup of the site in any way. Instead, it moved area residents from a position of vehement opposition to one of nonopposition or acceptance by placing within their grasp the opportunity to confirm for themselves that the emissions in the neighborhood were not excessive.

Accommodations in regard to traffic and hours of operation are also common in conjunction with the cleanup of contaminated sites and the routine operation of many facilities. Facilities or sites that have near neighbors may curtail certain early morning, weekend, or evening operations to avoid excessive noise. Truck traffic may be directed down certain streets so that it doesn't go past grade schools or through residential areas. Most of these types of accommodations are made as a result of residents' suggestions or complaints, and they can serve as important confirmation of organizational responsiveness to community wishes.

Sharing monitoring or test data with stakeholders can also be an important accommodation. In many cases, monitoring data or other test data under government programs may be available through the Freedom of Information Act. However, when organizations provide this information themselves—promptly—and provide stakeholders assistance in interpreting it, then the sharing of such information can rightly be called an accommodation.

Direct Stakeholder Involvement in Decision Making

Most of the situations we have discussed in this book have involved communication and soliciting of information from stakeholders, notably "the public," to "inform" or provide input into the decision-making process rather than allowing public stakeholders to make binding decisions themselves. Relatively little formal research has been done regarding what happens when community stakeholders are invited into active environmental decision-making processes. The role played by direct and active public input into the decision-making processes surrounding specific environmental concerns is still relatively new. There are certainly many instances in which stakeholders who have held veto power over siting or zoning decisions have chosen to deny requests for facilities or projects. However, there is also anecdotal evidence regarding many successes in the siting and permitting of various types of facilities over the past decade that have been achieved through the techniques we have discussed in this book—community assessments to identify stakeholders, involving stakeholders in a dialogue, thorough sharing of information, some accommodations to meet local needs or allay concerns, and so forth. In these cases, communities that could have said "no" said "yes," apparently because they had been provided with levels of knowledge and empowerment necessary to feel comfortable with accepting the facilities.

Thus, some of this anecdotal evidence indicates that providing stakeholders with an active role in the decision-making process *can* indeed lead to acceptance of facilities or projects that might have been rejected outright if the communities had been less involved in the decision-making processes. In some tough situations, at least, giving residents a say gives them a stake and a vested interest in the outcome that can make all the difference.

The following case describes one such situation involving a cleanup that suggests that, given the opportunity for greater control, residents may be willing to make decisions that those who are skeptical about community involvement in decision making would find surprising.

A small town in a rural area was faced with the cleanup of several contaminated lagoons. Several remedial alternatives were identified to clean up the site. One of the alternatives involved digging up all of the soil and sludge and transporting it to a landfill. Another alternative involved operation of a pump-and-treat system to clean up the groundwater and bioremediation to clean up the soil.

The first option was the most expensive by far, and involved "exporting" the contaminated material to a landfill in another community; however, it would also have removed the contamination right away.

The pump-and-treat/bioremediation option was far less costly, but would necessarily take a number of years to complete, thus leaving contamination in the community for some time.

The state agency involved in the cleanup went to the community well ahead of the hearings that would be held regarding the remedial alternatives and presented the information to residents. The residents were asked to consider the information and come up with their own recommendations regarding what they would like to see happen in their community.

Interestingly, community residents chose the pump-and-treat/bioremediation option. They had carefully reviewed the information during a series of town meetings that they held themselves and, although they didn't like having the contamination in their community, they believed that, based on the agency's risk assessment data, they would not be exposed to an unacceptable level of risk from the use of this option. The residents also chose the pump-and-treat/ bioremediation option because they decided that they did not wish to export a problem to another community if it could be handled adequately within their own borders.

This outcome says a lot about the inclusiveness of this community's decision-making process, as well as the excellent job that the agency did in supplying information and supporting the decision-making process.

It is likely that, as greater input into environmental performance of both public and private sector organizations is required by statute or through regulation, direct and active involvement of the public in the decision-making process will become more frequent. It will be interesting, as this practice evolves, to see what effect it will have on the quality of decision making and on community relations processes.

And If, After All of This, Stakeholders Still Object?

In the case of environmental issues, some stakeholders may remain opposed no matter what an organization does. As we discussed earlier in this chapter, moving critical stakeholders to a position of nonopposition is often the best outcome that can be obtained. If the majority of stakeholders support or hold positions of nonopposition toward the facility, site, or project, then the organization's goals can usually be attained (e.g., permit granted, expansion approved, remedial alternative adopted).

Unfortunately, in some situations, stakeholder concerns or opposition do not seem to be amenable to resolution through a community relations dialogue and the organization cannot achieve its goals. Serious differences of opinion or belief, too much "bad blood" or bad experiences, or the presence of too many stakeholders who are too angry or are pursuing other, unrelated agendas may overwhelm even the best conceived and best executed community relations program. In such cases, the organization will typically make a "business decision" to bow to community concerns.

ADDITIONAL CONSIDERATIONS IN IMPLEMENTING A COMMUNITY RELATIONS PROGRAM

The following topics have been touched on in previous sections of this book. The authors believe they bear repeating, however, especially within the context of community relations program implementation. Each of these topics goes to the heart of a solid community relations program.

Involving Employees

Employees are both stakeholders in their own right and important sources of information to other community residents. Community relations programs should consider employees as stakeholders who need information and can provide valuable input. Conversely, the role that employees can play in the community relations effort should also be considered.

Employees who are residents of the community in which a facility, site, or project is located are well positioned to speak informally to neighbors and friends about issues or operations, and should be well informed so they can speak authoritatively, since what they say is likely to be taken as fact.

In the case of one facility, employee comments about "lots of small spills" kicked off considerable controversy after the facility manager announced that the plant had had only three spills in five years. The facility manager was referring to reportable spills; the employees, who did not know the regulations, were referring to spills of less than reportable quantities, all of which also took place within secondary containment.

Neighbors believed the employees, since they were neighbors too, and thought that the facility manager was attempting to mislead the community.

As we discussed in chapter 4, employee involvement in "maintenance" activities in communities, such as speakers bureaus, cleanup days, or youth sports, can certainly help build and maintain positive relationships in the community. Additionally, in some communities (or among some stakeholder groups) in which residents shy away from management-level personnel, knowledgeable "line" personnel can help add legitimacy to the communication effort.

The Personal Touch

One of the more important concepts that we have endeavored to convey in this book is the importance of personal communication. There is no substitute for personal communication in building relationships or establishing trust. Although fact sheets can communicate concrete data and background information, they cannot convey empathy, interest, or understanding. Questionnaires can gather information on people's concerns, but they cannot pick up the little side comments, the shrugs, the shake of the head. Personal communication is frequently the catalyst needed to convince stakeholders that the information provided to them is valid and that their concerns are being heard and given consideration in the decision-making process.

The manager of the Illinois Environmental Protection Agency's (IEPA's) Office of Community Relations, Greg Michaud, conveys an interesting and very telling anecdote about the power of personal involvement in the communication process:

A residential neighborhood was undergoing cleanup of PCB contamination. The contamination necessitated soil being removed from people's yards. IEPA's community relations staff had been working with community residents to help them understand the cleanup process and the risk from any residual contamination left after cleanup. As would be expected, fear and a certain amount of distrust were issues that agency staff had to deal with.

During the course of the community relations effort, it became apparent that one individual in particular was an important opinion leader. He rarely said much to IEPA's community relations staff, but it was very clear that residents listened to him.

During a visit to the area, Mr. Michaud was asked by one of the residents to explain what a PCB was. While a group of residents—including the opinion leader—looked on, Mr. Michaud drew a rough diagram of a PCB molecule in the dirt with a stick, and explained about its chemical structure and use.

"After I did that, I was OK," says Mr. Michaud. "Something about the personal level of the communication said something to them—and to this man who was a leader in the community. After that, the communication just opened up."

Mr. Michaud's anecdote is one that community relations specialists can relate to—the moment when the dialogue clicks and, regardless of whether some differences in belief or opinion still exist, a genuine and positive relationship is engendered.

The Importance of Maintaining Good Environmental Performance

Activists sometimes criticize community relations, describing it as a shield used to make poor environmental performance acceptable to the community. The authors contend that community relations cannot be effective in the long run if the organization responsible for a facility, site, or project is not also working hard at improving or maintaining its environmental performance.

Even the best communication and the closest relationships cannot (and should not) substitute for good environmental performance. It is not enough simply to talk about a commitment to environmental protection. Stakeholders also have to see evidence that the organization in question is improving or maintaining its environmental performance at a high level. Substituting talk for good performance will ultimately lead to considerable anger on the part of community residents. As we have discussed, this anger, based on residents' belief that they have been betrayed and misled, is likely to eclipse the anger over the actual environmental performance.

Does the public expect perfection? Certainly, residents would prefer less (or no) environmental impact over more environmental impact, but most stakeholders are aware that emissions occur, natural resources must be harvested, and incidents with some environmental effects may occur. What people do expect is that public and private sector organizations will follow the laws and regulations, and that they will make efforts to safeguard human health and the environment. By improving or maintaining good performance and by maintaining an open dialogue with stakeholders about what they are doing, organizations can typically garner the support or nonopposition of most stakeholders.

CASES IN IMPLEMENTING COMMUNITY RELATIONS PROGRAMS

A number of organizations, both public sector and private sector, have implemented community relations programs. Much of the information in this book was developed based on their experiences of what works and what doesn't when dealing with public concerns regarding environmental issues.

In this section, we take a brief look at the programs designed to address public concerns about environmental issues implemented by four organizations: the Illinois Environmental Protection Agency, The Dow Chemical Company's Midland, Michigan, plant; Safety-Kleen Corporation; and the Chemical Manufacturers Association.

IEPA's Office of Community Relations: Communicating Early and Often

The Illinois Environmental Protection Agency (IEPA) includes a dynamic Office of Community Relations. Over the past 20 years, the Community Relations staff has been involved in working with the public at about 800 sites. At sites where effective public involvement has been undertaken, no litigation has interrupted or halted cleanup activities. Conversely, during the same time period, cleanup was slowed or stopped at four other sites in the state where effective public dialogues had not been implemented.

IEPA's Office of Community Relations comprises a staff of 11. The office's community relations/public involvement specialists are required to possess both technical knowledge and communication skills to ensure that they can engage in a substantive communication effort about environmental issues. Additionally, prospective employees must have a solid working knowledge of risk communication and the principles set forth by risk communication experts Peter Sandman and Vince Covello, since, as the office's manager, Greg R. Michaud, says, "Chances are much improved that our staff will be able to deal with emotional issues successfully if they have an understanding of risk communication."

Communicating Early and Often Much of the Office of Community Relations' success can be placed on its philosophy of "communicating early and often." In the case of site investigations and cleanups, for example, Mr. Michaud explains, "We believe that we have to start as soon as possible."

Unlike the U.S. EPA, which, in keeping with federal guidances for community relations, typically does not proceed with formal public involvement activities until the remedial investigation phase of the investigation and cleanup, IEPA begins its outreach activities as soon as a site is identified. This allows the agency's community relations specialists to build relationships in the community three or four years before a site is listed under CERCLA. Says Mr. Michaud, "We prefer to get started early. Then we don't have residents asking us, 'Where were you four years ago? Why didn't you tell us about the contamination when you first learned about it?'"

Additionally, IEPA's Office of Community Relations focuses much of its attention on small group meetings and one-on-one sessions, rather than on major public meetings. Although such large group meetings are part of the process, the agency's public involvement specialists prefer to hold small group meetings and one-on-one sessions that are more conducive to answering questions and dealing with issues beforehand.

"In the case of environmental and health concerns, people need to go through the process of dealing with the issues—the cycle of grief, understanding, acceptance—in order to move on. We can help them through this cycle a lot better in a small group setting," says Mr. Michaud.

Use of Community Assessments The Office of Community Relations uses the community assessment process to design effective approaches for each site that it handles.

"A doctor wouldn't perform an operation without conducting tests, and an engineer wouldn't build a bridge without a design. I'm amazed when people don't see why they should perform a community assessment before they launch a public involvement program," says Mr. Michaud. The assessment process is used to identify who is affected by a site, as well as those who think they are affected. A community assessment will make a public involvement program more effective and help reduce costs.

IEPA's public involvement specialists also identify the issues and concerns that stakeholders might have about the issue in question. Mr. Michaud notes that this includes what stakeholders *think* the issues are, as well as what the agency *believes* that they are. "What we think the issues are and what the public thinks they are can differ considerably," says Mr. Michaud, adding that issues that may be dubbed as "irrelevant" by technical staff can create major stumbling blocks. "If we can't deal with those issues, we can't deal with the big issues," explains Mr. Michaud, saying that spending an extra few minutes to address the seemingly insignificant issues always pays off in the long run.

Office of Community Relations staff also pay considerable attention to determining how best to provide information to stakeholders. "In some communities, we may find that residents listen to the radio or read a small weekly paper. Placing notices in a major daily newspaper might not reach the people we need to reach," says Mr. Michaud. Additionally, the community relations specialists identify local opinion leaders who they can work with to reach residents and establish solid links with the community. This enhances informal communication, and increases the likelihood that even the "irrelevant issues" that can, in fact, create major problems are dealt with.

Determining What Works, What Doesn't Since public involvement is still an evolving discipline, the Office of Community Relations conducts surveys on the effectiveness of the public involvement and communication efforts at some of its sites. Much of these survey data are collected on communication tools, such as fact sheets, to ascertain what residents liked or didn't like about the approach that was used. "When we have the time, we also like to go back and talk one-on-one with some of the residents when the TV cameras and the reporters are gone, to get additional feedback," says Mr. Michaud.

The Office of Community Relations also works with private companies, both in regard to permitting issues and cleanups, and in conjunction with other IEPA programs, such as the agency's Partners in Pollution Prevention (PIPP). Companies that sign on to PIPP are encouraged to undertake voluntary community outreach activities to communicate about their pollution prevention initiatives—an aspect of the state's pollution prevention program that sets it apart from the programs in most other states. The Office of Community Relations encourages companies to take a broad approach when it comes to permitting activities and other environmental issues. Explains Mr. Michaud, "How can you go into a community to talk about a water permit and not look at air issues and land issues?"

Some of the triumphs the Office of Community Relations has had have involved working with companies that have decided to work with IEPA on a voluntary basis to stave off public concerns over environmental issues. In one such case, a facility had lost a large quantity of product from an underground storage tank. A number of private water wells serving a nearby housing development lay directly in the path of the groundwater that had been contaminated by the release. Not only did the area boast a very active—and vocal—environmental organization, but the company that owned the facility was, at that time, also seeking to close a deal with some outside investors who wanted to invest in the plant.

"I sat down with the CEO and discussed with him what happens when a private well becomes contaminated. Then I explained what we could do to work with the public beforehand with proactive disclosure," says Mr. Michaud. Although the CEO was skeptical, he agreed to the course of action laid out by IEPA. "We talked to the neighbors and the representatives from the environmental organization and we explained how the cleanup would be conducted and how the groundwater monitored. Not only *didn't* the environmental organization use the situation to build its membership, but there was no press coverage, either," says Mr. Michaud.

The Dow Chemical Company: Community Outreach Program at the Midland, Michigan, Plant

The Dow Chemical Company's community outreach programs are very much a part of the company's corporate culture and its history. The company was founded in Midland, Michigan, by Herbert Henry Dow in 1897. One hundred years later, The Dow Chemical Company, a diversified manufacturer and marketer of basic chemicals, plastics, and specialty products and services, is the sixth largest chemical company in the world and the second largest in the United States.

The importance placed by corporate management on good relations with the Midland community—where the company's headquarters and one of its major manufacturing plants are located—has shaped the company's vision of community outreach at all of its plants worldwide. Although the actual form a community outreach program may take will vary from community to community to suit local needs, each of Dow's approximately 200 manufacturing sites around the world actively engages in community outreach.

In addition to plant-specific programs, another unique component of Dow's community outreach efforts is its Corporate Environmental Advisory Panel, which comprises representatives from all over the world. This panel provides worldwide perspectives to company executives on environmental issues from a number of international authorities.

Community Outreach at the Midland Plant Community outreach is especially important in Midland, where the company is by far the largest employer, with a tradition of philanthropy and community involvement dating from Mr. Dow himself. The community outreach program at the Midland plant involves a broad mix of activities, including a steady stream of one-on-one and small group meetings with a variety of stakeholders, an active community advisory panel, ongoing work with the media, a formal plant tour program, biannual perception surveys of area residents, and an educational outreach program. The plant airs information on the company and current issues to employees through an in-plant television system; this programming is also aired on a local cable channel. About 10 percent of the area residents report having seen the cable show.

A strong environmental element permeates most of the communication in the Midland plant's outreach program. The plant is proactive in issuing information on its environmental performance—whether releasing "good news," such as declines in emissions figures, or "bad news," such as permit exceedences.

"Because of the perception of chemicals, we need to have acceptance in the community," says Cindy Newman, manager of public relations for the Midland plant, who is responsible for the plant's community outreach program. A member of the Public Affairs Department, Ms. Newman is also responsible for public relations and media relations, and works closely with the Employee Communications manager in regard to Midland plant news and activities. As Ms. Newman points out, these varied duties frequently overlap—since issues involving the media frequently have a community outreach and employee communication component—and community outreach issues may require working with the media.

According to Ms. Newman, the vision driving Dow's community outreach program is the desire to ensure that the company's neighbors "welcome Dow into their communities" wherever the company has facilities. The objectives and goals used to achieve this vision include developing positive relationships with community leaders and putting in place processes and systems to ensure that outside perspectives are consistently conveyed to the company and other strategic partners for use in decision making.

Key Community Outreach Activities at the Midland Plant In addition to handling communication on all manner of environmental issues as they arise, the community outreach program at the Midland plant includes several major, ongoing activities designed to enhance the plant's dialogue with the community. These activities are described in the following paragraphs.

Community Advisory Panel　The Midland plant established its community advisory panel (CAP) in 1990 as part of the facility's commitment to the Chemical Manufacturers Association's Responsible Care® initiative. The Midland CAP is an independent group of 20 Midland area residents, which meets monthly with management from the company's Michigan division to discuss matters of interest to Dow and the community. The CAP, which is professionally facilitated, does not formally approve or disapprove company actions; instead, it provides the community's viewpoints. Although the meetings are closed to the public, the minutes of the meetings are available for review at the public library.

In response to CAP members' desires to ensure that an even wider range of perspectives be considered by the company, the Midland plant initiated a new program, "Extended Community Outreach Through the Community Advisory Panel," in 1994-95.

Extended Community Outreach Through the Community Advisory Panel　The Midland facility holds a "State of the Community" meeting once a year, in which the company invites some 300 community leaders to hear presentations about the plant, the company, and its operations, and to ask questions. To encourage additional input, breakout sessions have been added to the meeting for the past two years. During these 45-minute breakout sessions, small groups convene with a CAP member, a facilitator, and a Dow management representative to discuss those issues that they wish to convey to Dow management. These comments are recorded by the facilitator on a flip chart.

These comments have become the basis for the CAP's program of work. CAP members then work with the company to determine how to address the issues. Interestingly, the top issues that surfaced during the last round of breakout sessions did not focus on environmental issues. Instead, they focused on human relations and human resource issues related to the plant and the role that community residents perceived it as playing in the community.

The "Extended Community Outreach Through the Community Advisory Panel" program was awarded an honorable mention in Dow's internal 1995-96 Environmental Care Awards.

Community Perception Study At the suggestion of the CAP, Dow began conducting biannual surveys of residents' perceptions of the company in 1991. The surveys were initiated to establish a baseline and subsequent measures of residents' perceptions of Dow and its performance in Midland. Thus far, three surveys have been performed. The company intends to conduct the series of surveys over a 10-year period. The surveys, which are conducted by an opinion research firm by telephone, gather information on residents' opinions of the company in the areas of the environment, communication, and community impact. Three hundred Midland residents and 50 Freeland residents (Freeland is a nearby town downriver of the Midland plant site) are chosen at random to participate in the survey. Results of the surveys are used by Dow management and the CAP to determine what actions the company might need to take to respond to local perceptions.

Tours of the Midland Plant Tours are an important part of the community outreach program at the Midland plant. Fourteen years ago, the company built a visitors' center, which includes displays and exhibits, and launched a formal tour program to demystify its operations. Approximately 7,000 visitors a year—three-quarters of them school children from all over the state—tour the Midland plant to observe such processes as the making of SARAN WRAP® brand plastic film,[2] one of the products manufactured in Midland. Visitors board tour buses at the visitors' center and are escorted around the facility by retired Dow employees or former school teachers. Five to eight tours, on average, are run each week. Ms. Newman adds that the tours are so popular that the plant has, at times, had to turn people away on the more popular dates.

All of the tours include substantial information on the environmental aspects of the facility, since, as Ms. Newman says, "That's the message we want them to take away."

Although almost half of the respondents to the 1995 community perception survey of area residents indicated that they had been on a plant tour, says Ms. Newman, "That's low, we want everyone to tour our plant," adding that the survey has found that those who tour the plant have a more favorable view of the company and its environmental performance.

---

2. Trademark of The Dow Chemical Company.

Educational Outreach Involvement in the schools is another major, ongoing component in Dow's community outreach program. "As a company based on science, we know we need to do all we can to get kids turned onto science at an early age," says Ms. Newman. Educational outreach at the Midland plant includes teacher workshops, which allow teachers to visit plant laboratories and learn more about chemistry; sponsorship of minority student science programs; and, of course, tours. "With some teachers," says Ms. Newman, "the tour is a thing they do with their class each year."

Safety-Kleen Corporation: Implementing a Comprehensive Community Relations Program

Safety-Kleen Corp. is the world's largest recycler of hazardous and nonhazardous automotive and industrial fluids, serving approximately 400,000 customers worldwide. The company has more than 200 government-permitted facilities, located in 200 communities throughout North America. These include 14 "Recycle Centers," large facilities where the actual recycling/reclamation of the collected solvents and other wastes takes place; three used oil processing facilities, including the world's largest used oil re-refinery; and numerous smaller "branch" location used to collect and store the wastes prior to shipment to a Recycle Center or used oil processing facility. Most of the U.S. facilities possess RCRA operating permits that are broad in scope.

Because of the nature of Safety-Kleen's business—the collection and recycling/reclamation of wastes, many of which are hazardous—the company has, of necessity, become intimately acquainted with the importance of working with local government and community stakeholders.

Says Paul Wyche, vice president of corporate public affairs, which oversees community relations at Safety-Kleen, "We have found that communities are empowering themselves. They are recognizing that they can say 'yes' or 'no' to activities within their borders—and they really don't have to give a reason why. Residents want businesses to understand that they have a right to ask us about what we are doing in their communities. And, you know what? They do have that right."

Recognizing the Need, Implementing the Program Safety-Kleen's formal community relations program was implemented in the early 1990s as the company's senior management recognized a need to work more closely with government officials and community residents. Earlier in the company's history, most of its facilities kept low profiles, with minimal involvement with community stakeholders. Says Mr. Wyche of the pre-community relations days, "Those people who did know of us basically knew only that we were that 'hazardous waste' company. So there we were, thinking we were the good guys in the white hats who were performing this vital service—recycling hazardous and nonhazardous wastes—but many of the community members who knew us didn't perceive us that way at all."

To Mr. Wyche, who was recruited in 1991 to launch the company's community relations effort, community relations is about building relationships. Says Mr. Wyche, "What matters is the relationships that you build to communicate to the public the value that your company or your facility provides to the community. Such relationships can be successful only if the communication is two-way—if we listen as well as talk—and if we acknowledge that our businesses are in communities because their residents have given us permission to be there."

Community Assessments and Media Training   The approach that Safety-Kleen used to implement its community relations program included two important initiatives, which continue to form the backbone of the company's outreach efforts. First, more than 400 of the company's managers were trained in media relations and crisis communication, to sensitize them to the public's point of view. Second, community assessments were performed in more than 100 facility communities to identify stakeholders' questions and concerns. The assessments were performed to allow senior management to identify the communities that needed immediate attention in regard to community relations.

Mr. Wyche explains that the goal of training the company's managers in media relations and crisis communication was not so much the mastery of specific skills, but more as a method for sensitizing the managers to the way the company is perceived through the eyes of the media and the public. Particularly in the case of crisis communication training, managers are made aware of the viewpoints of local community residents, government officials, and the media. "They all have a stake in the outcome of a situation and a perspective that must be respected and addressed," says Mr. Wyche.

The media relations and crisis communication training also drives home to company managers that community relations is a priority—that they not only have permission to speak to the public, but that they have the responsibility to communicate on behalf of the company. "Officials, media representatives, and residents should not be hearing about what's happening in their community from a corporate spokesperson located a thousand miles away," adds Mr. Wyche, "They need to hear from the manager in the local community."

In regard to community assessments, Mr. Wyche credits the process with providing important insights into key community issues and into how the company can build lasting relationships with key stakeholders in the community. Of the approximately 100 assessments performed since 1992, about two dozen have identified issues that the company has been able to address to minimize or eliminate questions or concerns about its operations. The information from the assessments has played a crucial role in designing programs specific to each community. These programs include both formal and informal communication and one-on-one meetings to build solid relationships.

Mr. Wyche emphasizes that the media relations/crisis communication training and community assessments have been a means to an end—the establishment of strong relationships maintained through a positive dialogue. Through these strengthened relationships and intensive communication efforts, Safety-Kleen has successfully handled a number of issues regarding permitting and environmental performance that might otherwise have led to serious concerns. The manner in which the community relations program was implemented has also led to an important cultural change in the company, as community relations activities are now being carried out at the facility level, with guidance and assistance from corporate management, rather than at its direction.

Employee Involvement: Community Relations Teams At this point in the evolution of Safety-Kleen's community relations program, multi-year community relations plans have been prepared for a number of the North American facilities, and Recycle Center and used oil processing facility managers' annual reviews include assessments of their performance in community relations. Additionally, since facility managers cannot "do it all," the Corporate Public Affairs department has begun implementing a team approach in all Recycle Centers and used oil processing facilities, as well as in dozens of the branches. Corporate Public Affairs personnel work with facility personnel to build "Community Relations Teams," or "CR Teams," which are typically composed of about five people, and can include managers, secretaries, sales representatives, and warehouse persons. Corporate Public Affairs staff work with the CR Team members to increase their sensitivity to community issues and enhance their skills. Says Mr. Wyche, "Not only does this approach take some of the burden off of the facility manager, it also has the added benefit of involving additional employees." The company has also established "Community Liaison Committees" at its Recycle Centers and used oil processing facilities and at several of its larger and more high-profile branch locations.

The Chemical Manufacturers Association's Responsible Care® Initiative— "A Public Commitment"

The Chemical Manufacturers Association's (CMA's) Responsible Care® initiative provides an integrated approach to effective community relations. "Responsible Care®, the industry's health, safety and environmental performance improvement initiative, has become the ethical framework around which CMA member companies are operating. Through Responsible Care®, members reach out and involve the public in their performance improvement activities,"[3] explains former CMA chairman of the board and current chairman and CEO of Eastman Chemical Company, Earnest W. Deavenport, Jr.

Not only does the initiative require the approximately 190 CMA member companies—which represent roughly 90 percent of the chemical manufacturing capacity in the United States—to engage in proactive dialogues with community residents and other stakeholders, including employees, suppliers, customers, and local emergency responders, but it also provides a framework for continuous improvement in regard to overall environmental performance. "The Guiding Principles of Responsible Care®," according to which CMA member companies pledge to manage their businesses, appear in the box on p. 246.

CMA's Responsible Care® initiative, which it refers to as "a public commitment," is a particularly powerful program for community relations because it focuses extensively on performance, as well as on the needs of a wide variety of stakeholders.

Responsible Care® comprises six Codes of Management Practices, which include the following:

- The Community Awareness and Emergency Response (CAER) Code, which promotes an active dialogue with facility communities and local emergency planning.
- The Pollution Prevention Code, which commits member companies to the safe management and reduction of wastes.

---

3. Chemical Manufacturers Association, *Responsible Care® In Action, 1993-94 Progress Report*, p. 2.

Guiding Principles of Responsible Care®

■ To recognize and respond to community concerns about chemicals and our operations.

■ To develop and produce chemicals that can be manufactured, transported, used and disposed of safely.

■ To make health, safety and environmental considerations a priority for our planning for all existing and new products and processes.

■ To report promptly to officials, employees, customers and the public, information on chemical-related health or environmental hazards and to recommend protective measures.

■ To counsel customers on the safe use, transportation and disposal of chemical products.

■ To operate our plants and facilities in a manner that protects the environment and the health and safety of our employees and the public.

■ To extend knowledge by conducting or supporting research on the health, safety and environmental effects of our products, processes and waste materials.

■ To work with others to resolve problems created by past handling and disposal of hazardous substances.

■ To participate with government and others in creating responsible laws, regulations and standards to safeguard the community, workplace and environment.

■ To promote the principles and practices of Responsible Care® by sharing experiences and offering assistance to others who produce, handle, use, transport or dispose of chemicals.

Reprinted with permission of the Chemical Manufacturers Association.

■ The Process Safety Code, which is designed to prevent accidental chemical releases.

■ The Distribution Code, which minimizes risk from the transportation, storage, transfer, and handling of chemicals in transit.

■ The Employee Health and Safety Code, which protects the health and safety of employees and visitors to member companies' plants.

■ The Product Stewardship Code, which manages chemicals from research and design through recycling and disposal to lessen adverse impact on the environment.

Although the CAER Code addresses outreach in depth, each of the other codes also includes an outreach component to various stakeholder groups among its management practices.

The Community Awareness and Emergency Response Code

Community Awareness and Emergency Response (CAER) was first initiated in 1985 in the wake of the Bhopal, India, accident to promote coordinated emergency response planning in plant communities, encourage more openness between the chemical industry and the communities in which plants were located, and instill confidence in the chemical industry among the general public. The CAER initiative preceded the passage of the Emergency Planning and Community Right-to-Know Act (EPCRA) (Title III of the Superfund Amendments Reauthorization Act of 1986), which mandated the establishment of state emergency response commissions (SERCs) and local emergency planning committees (LEPCs). Although the emphasis of the early CAER initiative was on emergency response, CMA also encouraged its member companies to engage in dialogues with local officials, residents, and other stakeholders in their communities.

In the late 1980s, CMA's board recognized a need for a fully integrated approach to both communication and environmental performance and implemented the Responsible Care® initiative based on the framework of a similar program that had been initiated by the Canadian chemical industry. CAER, with increased emphasis on community outreach, became the first of Responsible Care®'s Codes of Management Practices. The following lists the 19 Management Practices under the CAER Code.

Community Awareness and Emergency Response Code Management Practices*

A. Community Awareness and Outreach

   Member facilities that manufacture, process, use, distribute or store hazardous materials shall have a community outreach program that includes:

   For Employees:

   1. An ongoing assessment of employee questions and concerns about the facility.

*Reprinted with permission of the Chemical Manufacturers Association.

2. Communications training for key facility and company personnel who communicate with employees and the public concerning safety, health, and environmental issues.
3. Education of employees about the facility's emergency response plan and safety, health, and environmental programs.
4. An ongoing dialogue with employees to respond to their questions and concerns and involve them in community outreach efforts.
5. A regular evaluation of the effectiveness of the ongoing employee communications effort.

For the Community:
6. An ongoing assessment of community questions and concerns about the facility.
7. An outreach program to educate responders, government officials, the media, other businesses and the community about the facility's emergency response program and risks to the community associated with the facility.
8. A continuing dialogue with local citizens to respond to questions and concerns about safety, health, and the environment, and to address other issues of interest to the community.
9. A policy of openness that provides convenient ways for interested persons to become familiar with the facility, its operations, and products, and its efforts to protect safety, health, and the environment.
10. A regular evaluation of the effectiveness of the ongoing community communications efforts.

B. Emergency Response and Preparedness
   Member facilities that manufacture, process, use, distribute, or store hazardous materials shall have an emergency response program that includes:

1. An ongoing assessment of potential risks to employees and local communities resulting from accidents or other emergencies.
2. A current, written facility emergency response plan which addresses, among other things, communications and the recovery needs of the community after an emergency.
3. An ongoing training program for those employees who have response or communications responsibilities in the event of an emergency.
4. Emergency exercises, at least annually, to test operability of the written emergency response plan.
5. Communication of relevant and useful emergency response planning information to the Local Emergency Planning Committee.

6. Facility tours for emergency responders to promote emergency preparedness and to provide current knowledge of facility operations.
7. Coordination of the written facility emergency response plan with the comprehensive community emergency response plan and other facilities. If no plan exists, the facility should initiate community efforts to create a plan.
8. Participation in the community emergency response planning process to develop and periodically test the comprehensive community emergency response plan developed by the Local Emergency Planning Committee.
9. Sharing of information and experience relating to emergency response planning, exercises, and the handling of incidents with other facilities in the community.

Evaluation and Public Input Managers of CMA member companies are required to perform an annual self-evaluation of these management practices to determine how far along each facility is in implementing them. As part of the implementation of the CAER Code, many CMA member companies have established community advisory panels (CAPs) at their facilities to receive input from community stakeholders. In addition, management at many facilities have initiated outreach activities, such as tours, open houses, community meetings, and meetings with neighbors, as well as engaging in involvement in local schools, to establish and maintain positive dialogues with their communities.

Important input into the Responsible Care® initiative comes from CMA's National Public Advisory Panel, a group of about 15 persons drawn from a wide variety of constituencies and representing different parts of the United States. The panel, which was first convened in 1989, has advised CMA on the development of the codes and has helped broaden CMA's understanding of public concerns regarding the manufacture and use of chemicals. CMA also conducts extensive research into the opinions and concerns of chemical industry employees, residents of plant communities, local emergency planning officials (LEPC members), and the general public to identify important issues and measure the success of member companies' communication efforts.

CHAPTER

7

REGULATORY REQUIREMENTS FOR ENVIRONMENTAL COMMUNITY RELATIONS

Recent federal regulatory actions demonstrate how public involvement in environmentally based legislation and decision making is on the rise. Facilities seeking to obtain certain types of permits must first notify the public and then provide an opportunity for public comment and scrutiny of the draft plans or permits. In many cases, the burden lies with both the facility and the regulatory agency to inform the public of its intentions and to provide it with the opportunity to participate in the decision-making process. Public involvement is also an important component of investigation and remedial design activities under CERCLA.

This chapter presents the public involvement and community relations requirements found in the Resource Conservation and Recovery Act (RCRA), the Comprehensive Environmental Response, Compensation, and Liability Act (CERCLA), and the Clean Water Act (CWA). Reporting requirements in the Emergency Planning and Community Right-to-Know Act (EPCRA) and the recent Risk Management Program regulations promulgated under the Clean Air Act (CAA), which also create a need for community relations activities to explain the reported information to the public, are also discussed.

REQUIREMENTS UNDER RCRA

The Resource Conservation and Recovery Act (RCRA) public involvement requirements are contained in Title 40 of the Code of Federal Regulations (40 CFR) Parts 124 and 270 and summarized in various EPA manual publications and guidances. The EPA primarily uses the phrase "public involvement" rather than "community relations," "public participation," or other similar terms when referring to RCRA activities in the regulatory guidances. The following discussion presents the essential components of RCRA public involvement as contained within the guidances and applicable regulations. First, a brief overview describing the major provisions of the act is provided.

RCRA Provisions

The Resource Conservation and Recovery Act was enacted in 1976, as an amendment to the Solid Waste Disposal Act, in order to address the increasing nationwide problem concerning the disposal of solid wastes. The intent of the RCRA legislation was to reduce or eliminate the generation and consequent disposal of hazardous wastes to as great a degree as possible. The act has continued to evolve throughout the years, perhaps most significantly with the passage of the Hazardous and Solid Waste Amendments (HSWA) in 1984. These amendments significantly expanded the scope of RCRA and resulted in the creation of corrective action provisions intended to ensure the timely identification, evaluation, and remediation of contaminated RCRA facilities.

Subtitle C of RCRA, promulgated in 40 CFR Parts 261-266 and Parts 268-270, established a program to manage hazardous wastes "from cradle to grave." The regulations identify the characteristics that are used to define a waste as "hazardous." The provisions regulate the generation, transportation, treatment, storage, and disposal of these hazardous wastes. RCRA regulatory requirements for treatment, storage, and disposal facilities (commonly referred to as TSDFs) comprise the largest category of provisions. In addition, the regulations set technical standards for the design and operation of hazardous waste facilities. Permitting requirements for all types of RCRA facilities are also contained in these regulations.

Following is a summary of the RCRA permitting and corrective action programs, both of which require public involvement activities throughout the regulatory process.

RCRA Permitting

Owners and/or operators of a facility that is regulated under RCRA must complete and submit a permit application to the regulatory agency that covers all aspects of the facility's operation. The permit application is divided into parts A and B. Part A is a standard form that simply provides for the submission of general facility information. Part B requires the provision of detailed information on the facility's design, operation, maintenance, record-keeping, and closure plans. There is no "standard form" for a Part B permit application, so the owner or operator must rely on the regulations for guidance (40 CFR Parts 264 and 270). Existing facilities that received hazardous wastes on or after November 19, 1980, applied for "interim status" with the submittal of a Part A application and then were required to submit a Part B application to achieve permanent operating status. The owner or operator of a new facility must submit Parts A and B of the application simultaneously at least 180 days before the date on which construction of the hazardous waste facility commences.

RCRA Corrective Action Program

Corrective actions are required at facilities where past or present practices caused a release of hazardous wastes or hazardous constituents resulting in contamination of the water or soil. The Hazardous and Solid Waste Amendments provided three corrective action provisions that expanded the EPA's authority to initiate such actions.

1. Section 3004(u) requires that any permit issued to a facility after November 8, 1984, under Section 3005(c) of RCRA address corrective action for releases of hazardous wastes or constituents from any solid waste management unit (SWMU) at the facility.
2. Section 3004(v) authorizes the EPA to require that the facility provide corrective action beyond its physical boundaries, if necessary.
3. Section 3008(h) authorizes the EPA to issue administrative orders or initiate court actions to require corrective actions when there has been a release at a RCRA facility operating under interim status.

The RCRA Part B permit or administrative order (in the case of an interim status facility) specifies the conditions under which the facility owner or operator must provide corrective actions. The EPA may also issue what is termed a *permit schedule of compliance*, which requires the owner or operator to begin or continue corrective actions in cases where the facility is issued a permit before the corrective actions are commenced or completed.

Expansion of Public Involvement

On December 11, 1995, the EPA issued a final rule to expand RCRA public participation provisions. The rule amends 40 CFR Parts 124 and 270 to increase the opportunities for public involvement by allowing public participation at an earlier point in the permitting process. These new regulations are incorporated into the following discussion of RCRA public involvement requirements during the permitting process.

RCRA Processes Requiring Public Involvement

The EPA requires that certain public involvement activities occur during the following RCRA processes:

■ As a precursor to permit applications
■ Review of permit applications
■ Issuance of permits and administrative orders
■ Permit modifications
■ Implementation of corrective action programs
■ Approval of closure plans

The goal of the required public involvement activities is to ensure that all interested persons and affected parties have the opportunity to participate in the decision-making process with regard to hazardous waste permitting and program management activities. The following sections address RCRA public involvement requirements and activities in the permitting process, for corrective action orders, and during closure.

Public Involvement in the RCRA Permitting Process

40 CFR Parts 124 and 270 contain the public involvement and activity provisions that the agency requires, beginning from a time prior to the initial permit application through final issuance of both operating and post-closure permits. The agency stresses that there are times that public involvement activities may be required to go beyond the specific provisions of Parts 124 and 270. Figure 7.1 displays the public involvement requirements during the RCRA permit decision-making process.

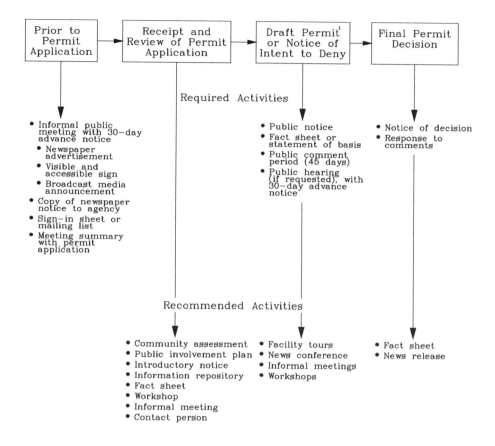

| Prior to Permit Application | Receipt and Review of Permit Application | Draft Permit[1] or Notice of Intent to Deny | Final Permit Decision |

Required Activities

• Informal public meeting with 30–day advance notice
• Newspaper advertisement
• Visible and accessible sign
• Broadcast media announcement
• Copy of newspaper notice to agency
• Sign–in sheet or mailing list
• Meeting summary with permit application

• Public notice
• Fact sheet or statement of basis
• Public comment period (45 days)
• Public hearing (if requested), with 30–day advance notice

• Notice of decision
• Response to comments

Recommended Activities

• Community assessment
• Public involvement plan
• Introductory notice
• Information repository
• Fact sheet
• Workshop
• Informal meeting
• Contact person

• Facility tours
• News conference
• Informal meetings
• Workshops

• Fact sheet
• News release

[1]The draft permit may include a schedule of compliance for corrective action, as well as conditions related to closure and post–closure care, if appropriate. These procedures are those followed by the EPA. When individual states are authorized to assume the lead, their procedures may vary.

■ **FIGURE 7.1 Public Involvement During the RCRA Permit Decision Process.**

Source: Adapted from Exhibit 3-1, *RCRA Public Involvement Manual*, U.S. Environmental Protection Agency, EPA 530-R-93-006, September 1993, p. 3-3.

Community residents may raise questions concerning facility operating conditions, closure, post-closure care, and corrective actions during the permitting activities. These potential questions and problem areas need to be identified as early in the process as possible. The permit applicant must hold at least one "informal" meeting with the public prior to submission of a Part B permit application. The same procedures hold in the case of a facility that is considering a change in operations that would qualify as a Class 3 permit modification (discussed in upcoming sections). The reason for the meeting is twofold. First, the meeting is intended to provide a forum for the facility owner or operator to inform the community about the proposed permit conditions and hazardous waste management practices. Second, the meeting is intended to provide community members with the opportunity to ask questions about the facility's proposed operations. The owner or operator must provide at least a 30-day advance public notice of the meeting through the following methods:

1. A newspaper advertisement. A display advertisement must be posted in a general circulation newspaper in the county or similar jurisdiction. The agency may require that the advertisement also appear in a newspaper serving an adjacent county or other location where potentially affected persons reside.
2. A visible and accessible sign. The notice must be posted on a clearly marked sign at or near the facility. If placed on the facility's property, the sign must be visible and readable from public access points.
3. A broadcast media announcement. A broadcast notice must be made at least once on at least one local radio or television station. The agency may waive this requirement if another acceptable medium is proposed.

The owner or operator must provide a copy of the newspaper notice to the agency and ensure that it includes the date, time, and location of the meeting; the purpose of the meeting; a description of the proposed operations and a map of the facility location; a statement encouraging 72 hours' advance notice for those persons needing special access assistance; and identification of a contact person.

The owner or operator must post a sign-in sheet or provide a similar opportunity for attendees to signify their desire to be placed on a mailing list. A summary of the meeting must be submitted to the agency as part of the permit application.

Permit Application Activities The agency requires the compilation of a mailing list at the time that the owner or operator submits the permit. The mailing list is composed of interested members of the public assembled so that necessary information about meetings, hearings, events, and available reports and documents can be effectively communicated to interested persons and affected parties. The agency typically compiles the list through consultation with local community organizations, neighborhood associations, local and state agencies, the facility, and other sources. The list usually includes local officials; interested, affected, and potentially affected community members; residential neighbors in proximity to the facility (usually within a one-half-mile radius); and media representatives.

Once the application has been submitted, the agency must publish a public notice "within a reasonable time" that includes the name and telephone number of the facility's contact person; the mailing address and contact person for the agency; the location where copies of the permit application and supporting documentation can be viewed; a description of the proposed application; and the date the application was submitted. The agency must place the permit application and supporting documents in a location that is convenient and accessible to members of the community who desire to review the information.

The actual need to establish an "information repository" is decided on a case-by-case basis. The agency considers the level of current or potential public interest, type of facility, presence of an existing repository, and proximity to the location of the nearest administrative record. If the agency determines that a repository is necessary, the facility owner or operator is responsible for establishing and maintaining it. The facility owner or operator must, at a minimum, provide written notice about the repository and its location to all persons on the mailing list.

Other public involvement activities may be initiated depending on the perceived or anticipated level of community interest in the facility. If the agency determines that the permit application is likely to draw a significant amount of interest, it may decide to conduct a community assessment and even develop a formal public involvement plan. A public involvement plan provides a community-specific guide for conducting communication activities in relation to the permitting or corrective action taking place at a RCRA facility. The plan is based on information collected during community interviews and through the community assessment. It focuses on assessing the level of community interest, responding to questions or concerns expressed by the community, and recommending specific activities for involving the community in the RCRA process.

Other activities that may be utilized include developing fact sheets and conducting small group meetings to keep the public informed about the status of the permit application. These activities are designed to be proactive in order to ensure that the affected public is included at an early stage in the decision-making process.

Activities During the Review Process   The agency's permit review process can be quite lengthy; it is not uncommon for more than a year to elapse between the initial permit application and final issuance. During that time, the agency may choose to undertake a number of activities to keep the community informed about the status of the permit. Workshops and informal small group meetings may be conducted to educate the community about the RCRA process and to keep the public interested and involved. A single agency contact person is typically assigned to provide continuity for the communications initiated and the public involvement activities performed. The agency attempts to garner as much community input as possible before the draft permit is issued, to ensure that it is responsive to the needs and concerns of the community and to increase the likelihood that the permit will be supported and accepted by the community.

Before issuing a draft permit decision, the agency may conduct a RCRA facility assessment (RFA) to identify all SWMUs and the potential for release of hazardous wastes or constituents at the facility. The RFA represents the first stage of the corrective action process and provides initial information on whether such actions are likely to be a significant issue for the facility. If information gathered during the RFA indicates that a release of hazardous wastes or constituents has occurred or has the potential to occur (based on site-specific circumstances) from an SWMU, the agency may require that the next step in the process, a RCRA facility investigation (RFI), be undertaken.

Public Involvement at the Draft Permit Stage   The agency must decide to either deny the permit application or prepare a draft permit for the facility at the conclusion of the review process. A draft permit must include technical requirements, possible corrective action schedules of compliance, and other conditions relating to the operational safety of the facility. If a draft permit is prepared for a facility, the agency must release it to the public for review and give formal public notice that it is available for comment. Similarly, the public must be notified if the agency plans to deny a permit application. Notice in both cases must be published in a major local newspaper and broadcast over a local radio station. Notice must also be sent to all persons on the mailing list.

A fact sheet must also be prepared in response to the draft permit or the notice of intent to deny the permit. The fact sheet must summarize the significant factual, methodological, policy, and legal foundations for the draft permit decision. The fact sheet typically includes a brief description of the facility and its operations, the basis for the draft permit conditions, and information about the required comment period and procedures to request a hearing. A *statement of basis* may also be prepared. This is similar to a fact sheet, although it typically does not contain the same level of detail and is often used in the case of permit denial.

A 45-day public comment period follows publication of the notice about the draft permit or intent to deny the permit. This 45-day time frame provides the public with the opportunity to comment, in writing, on the conditions contained in the draft permit or in the intent to deny the permit. The agency may extend the comment period or reopen the permit process if the comments raise significant questions about the draft permit. Additionally, commenters have the right to request a public hearing on the draft permit decision. If such a request is made in writing, the agency must plan and hold a public hearing with 30 days' advance notice to the community regarding the time, date, and location of the meeting. This is not to imply that the agency needs a request in order to schedule a public meeting. If the agency perceives that the community will have significant questions or concerns regarding a draft permit, it may initiate and plan a public meeting to coincide with the issuance of the decision.

Additional activities that can be implemented during the comment period depend on the level of interest expressed in the community about the facility. The agency may work with the owner or operator to provide a facility tour so that the public has an opportunity to see for itself how the operation functions. Workshops or small group meetings may be held by the agency or in conjunction with the facility to further educate the public about the facility's operations and permit conditions. Again, the agency may choose to conduct this type of activity earlier in the permitting process to gather as much information as possible about public concerns and questions.

Public Involvement upon Permit Issuance   The agency reviews all written and oral comments received and issues a final permit decision at the conclusion of the 45-day comment period. A *notice of decision* is sent from the agency to the facility owner or operator and to any persons who submitted public comments or requested the notice. A written *response to comments* is prepared by the agency that presents all significant written comments received during the public comment period, along with an explanation of how those comments were either addressed or rejected in regard to the final permit decision. The response to comments is intended to demonstrate to the community that its concerns and questions were considered in the decision-making process. All comments must be made available through the Administrative Record and in the information repository (if one was established).

The response to comments must be sent to the owner or operator of the facility and to all persons who submitted comments or requested a copy of the response. If the community expressed a high level of interest in the permit process, a fact sheet may be prepared to explain the decision to the public and sent to everyone on the mailing list. If the media expresses interest in the final permit decision, the agency may also prepare a news release for the major outlets to explain the final decision.

Permit Modification Public Involvement Activities   There are many reasons why a facility may need to modify a permit, such as process changes, new product development, or revised regulatory requirements. Public involvement is a component of the modification process; however, involvement responsibilities and activities vary, depending on whether the facility owner or operator or the agency initiates the modification. The level of involvement also depends on the degree to which the modification substantively changes the permit conditions. However, the only operating conditions open for public comment are those that are subject to the requested modification, regardless of who initiates the process.

Modifications that are initiated by the agency are subject to the same public involvement requirements and activities as described for the permitting process. Facility-initiated modifications are categorized as Class 1, 2, or 3, according to how substantively the permit is expected to change operating conditions. Facility-initiated permit modifications typically require that the owner or operator take on most of the responsibility for conducting public involvement activities.

Class 1 modifications.   These modifications address routine and administrative changes to the permit, such as updating, replacing, or relocating emergency equipment; schedule changes or updates; improving monitoring, inspection, record-keeping, or reporting procedures; and updating sampling and analytical methods. Such modifications do not substantively alter the permit's conditions. The only public involvement requirement for Class 1 modifications is that the facility notify the public, by sending a notice to all parties on the mailing list, within 90 days of implementing the change.

Class 2 modifications. These modifications address changes that must be made in order to respond effectively to the types and quantities of wastes present at the facility, technological advances, and new regulatory requirements. In order to be considered Class 2 modifications, these changes must be implemented without substantively altering the facility's design or management practices as described in the permit. The facility must submit the modification request along with supporting documentation to the agency.

The facility is required to notify persons on the mailing list about the modification request and to publish a notice in a major local newspaper of general circulation. The newspaper notice marks the beginning of a 60-day public comment period and must announce the date, time, and place of a public meeting. The notice must also identify a contact person for both the facility and agency and contain the statement, "The permittee's compliance history during the life of the permit being modified is available from the regulatory agency contact person."

The facility must place the request for modification and supporting documentation in an information repository or location accessible to the public in the vicinity of the facility. The facility must conduct the public meeting no earlier than 15 days after the start of the 60-day comment period and no later than 15 days before it ends. The meeting is intended to provide a medium for an exchange of information about the permit modification and an opportunity to address questions or concerns. The agency is not required to attend this meeting or respond to comments made during the meeting. However, the agency is required to consider all written comments submitted during the public comment period and must respond in writing to all significant comments in its final permit modification decision.

Class 2 modification provisions include a default action to ensure that the agency responds to the facility's request in a timely manner. The agency must respond to the written request within 90 days or, upon notification of an extension, within 120 days. If the agency does not issue a final decision within this time period, the facility is granted an automatic authorization to conduct the requested activities described in the written modification for 180 days. If the agency still has not acted within 250 days of receipt of the modification request, the facility must notify persons on the mailing list within seven days, and make a reasonable effort to notify other persons who submitted written comments, that the automatic authorization becomes permanent unless the agency approves or denies the request by day 300.

The agency must notify persons on the mailing list within 10 days of any decision to grant or deny a Class 2 modification request. The agency must also notify persons on the mailing list within 10 days after an automatic authorization for a Class 2 modification goes into effect.

The agency may also approve a temporary authorization to allow the facility, without prior public notice and comment, to conduct certain activities for up to 180 days, if it determines they are necessary to respond to changing conditions. The facility must notify all persons on the mailing list about the temporary authorization within seven days of the request. Temporary authorizations are typically issued to allow an owner or operator to perform a one-time only or short-term activity, for which the full permit modification process is deemed inappropriate.

Class 3 modifications. These modifications address changes that substantially alter a facility or its operations. The facility must submit a modification request and supporting documentation to the agency. Class 3 modifications are subject to the following same review and public participation procedures as defined for permit applications:

- Pre-application public meeting and notice
- Preparation of draft permit modification conditions or notice of intent to deny the modification
- Publication of notice of the agency's draft permit decision that establishes a 45-day comment period
- Development of a fact sheet or statement of basis
- A public hearing, if requested, with a 30-day advance notice
- Issuance of a notice of decision to grant or deny the modification
- Consideration and response to all significant written and oral comments received during the 45-day comment period
- Consideration and response to all significant written comments received during the 60-day public comment period

When public concern is perceived to be high, the agency may require that the facility perform additional public participation activities as described in the permitting process. These activities can include workshops, small group meetings, facility tours, and the preparation of additional fact sheets designed to educate the public about the intended change.

As with Class 2 modifications, the agency must notify persons on the mailing list within 10 days of a decision to grant or deny the Class 3 request. The agency may also grant temporary authorization to a facility to perform certain activities requested in the modification for up to 180 days without prior public notice and comment. However, the facility must issue a public notice to all persons on the mailing list within seven days of submitting the temporary authorization request.

Public Involvement in Closure and Post-Closure Activities

Any time an owner or operator seeks to close a solid waste management unit at a RCRA-regulated facility, certain procedures must be followed. Closure, in this context, involves completing treatment, storage, and/or disposal operations; application of final cover or cap to landfills; disposal of contaminated equipment or structures; and remediation of contaminated soil. The term post-closure applies only to land disposal facilities that have not achieved "clean closure." Clean closure is achieved when all contaminants are removed from the SWMU. Post-closure typically lasts for a 30-year period, during which monitoring and maintenance of the unit must be continued to ensure that contamination not removed during closure activities does not migrate out of the disposal system.

The public has the opportunity to comment on a facility's closure and post-closure plans and any amendments that are subsequently made to the plans during the permitting and modification processes. Regulations regarding the completion of closure and post-closure plans at permitted facilities are contained in 40 CFR Part 264.

Agency public involvement activities in connection with closure issues can include distributing fact sheets that describe the closure plans and process, along with conducting workshops or small group meetings. Particularly in cases where closure activities will mark the end of a facility's lifetime, the public may be very interested in the mechanisms that will ensure that the units are safely closed. In cases in which the owner or operator intends to leave the facility upon closure, the public may be concerned about whether adequate plans and financial assurances necessary to complete required closure and post-closure activities are in place.

Interim status facilities may also need to close certain units. Regulations contained in 40 CFR Part 265 require facilities seeking to close under interim

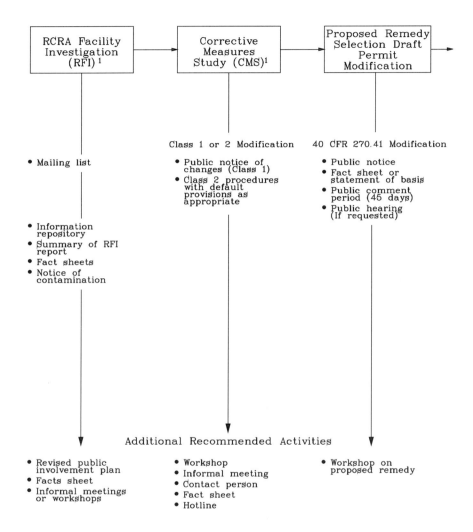

[1]The public has the opportunity for notice and comment on the permit schedule of compliance for corrective action, including its RFI and CMS provisions, at the time of permit issuance.

■ **FIGURE 7.2 Public Involvement Activities in the RCRA Corrective Action Process (during RCRA permitting).**

Source: Adapted from Exhibit 4-1, *RCRA Public Involvement Manual*, U.S. Environmental Protection Agency, EPA 530-R-93-006, September 1993, pp. 4-5.

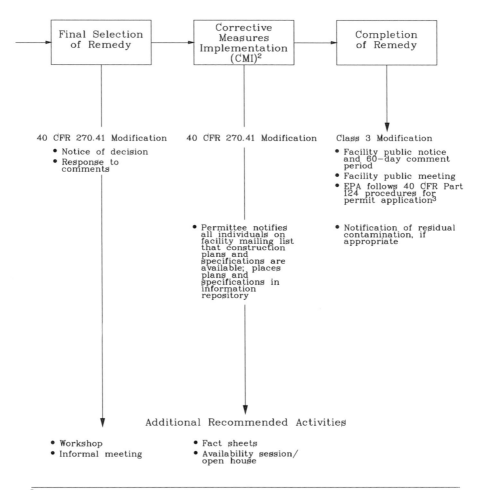

2As indicated in this figure, the public receives notice of and has the opportunity for comment on the permit schedule of compliance for corrective measures implementation during the remedy selection modification under 40 CFR 270.41.

3I.e., public notice, 45-day comment period, fact sheet or statement of basis, hearing if requested, notice of decision, response to comments submitted during both the 45-day and 60-day comment periods.

■ **FIGURE 7.2 Continued**

status to submit closure and post-closure plans. Public involvement activities required for interim status closures are also specified in these regulations. The regulations require that the agency provide the public, through a newspaper notice, with an opportunity to provide written comments on the closure and post-closure plans and request modifications to the plans no later than 30 days from the date of the notice. Commenters have the right to request that the agency hold a public hearing to discuss the planned closure activities. Public notice of the hearing must be provided at least 30 days prior to its scheduled date.

The public may also petition the agency to extend or reduce the post-closure care period applicable to the interim status facility or SWMU. If the agency is considering such a petition, it must provide the public and the facility, through a newspaper notice, with the opportunity to provide written comments within 30 days. Again, the public may petition for a hearing on the plan with the same notification requirements as described for closure.

Interim status facilities may also choose to amend their closure and post-closure plans according to specified requirements. If the amendment to either a closure or post-closure plan is a Class 2 or Class 3 modification, as specified earlier, the facility must follow identical procedures. Similarly, the same agency approval criteria and public involvement activities are required.

Corrective Action Public Involvement

A facility is usually brought into the corrective action process either when the agency is considering a permit application or through an enforcement order. The public may comment on the schedule of compliance for a corrective action during both the initial permit issuance and the permit modification processes.

As discussed, corrective actions are initiated when a release or the potential for a release of hazardous wastes or constituents from a solid waste management unit has been identified. Corrective action investigations can have a high profile and receive considerable attention from the public and the media. If the RCRA facility assessment (RFA) uncovers the potential for a release or the actual release of hazardous wastes, a number of steps may be taken within the corrective action process. These steps each require public involvement activities that are displayed in Figure 7.2, on pp. 264–65.

When a release or potential release at a facility has been identified through the RFA, the next step the agency takes is to conduct a RCRA facility investigation (RFI). The RFI, which may be conducted in phases, is undertaken to define and characterize the nature and extent of contamination at the facility.

The mailing list that was originally compiled during the permitting process is updated during the RFI to ensure that the necessary community members are informed about the activities. Additional community concerns that can arise during the investigatory activities may require that the public involvement plan, if one has been prepared, be revised. Fact sheets may be assembled and distributed throughout each phase of the RFI activities in order to keep the community informed about the progress of the investigation. A fact sheet should also be issued at the conclusion of the investigation in order to report the results. The agency also recommends that small group meetings and workshops be used to continue to keep the public updated on the investigatory process.

A proposed Subpart S rule (55 Fed. Reg. 3078, July 27, 1990) contains revised public involvement activities for corrective actions. Under the proposed rule, the agency can require the facility to establish an information repository in the community. The repository would provide a central location where all reports and data concerning the corrective action are made available to the public. Subpart S also proposes that the facility inform the public about the existence of the repository by sending a *notification of public information repository* to all persons on the mailing list. This notification must identify a contact person to whom comments can be submitted. The facility must publish this notice in a local newspaper and broadcast it on local radio or television stations.

Subpart S also proposes that the facility mail a summary of the final RFI report to all persons on the mailing list. In addition to the report summary, a fact sheet that explains the essential investigation results and conclusions would be required.

The proposed rule also requires the issuance of two types of notifications of contamination. The first notification would be issued upon the discovery of hazardous constituents in groundwater potentially resulting from a SWMU release and migrating beyond the facility's boundaries in concentrations exceeding risk-based levels. This written notification would be made within 15 days of discovery to both the agency and any person(s) owning or residing on land that overlies the contaminated groundwater. Similarly, the rule proposes a second notification to any individuals who have been, or may have been, exposed to air releases from SWMUs that could have migrated beyond the facility's borders in concentrations exceeding risk-based levels.

Corrective Measures Study A corrective measures study (CMS) is undertaken at the conclusion of the RFI activities to identify and evaluate possible alternatives to remediate contamination discovered at a facility. There are no specific additional public involvement activities required during the CMS beyond those associated with the initial permitting process or modifications to the corrective action schedule of compliance.

Because the length of time between the initial RFI and the CMS may extend to several years, however, additional public involvement activities are typically undertaken. Workshops and small group meetings are recommended to keep the public informed about the activities being conducted and the status of the investigation. Fact sheets are often developed and sent out at significant milestones during the study to keep the public up to date about its progress. The agency also recommends that a contact person be identified to answer questions from individuals in the community and to provide a consistent point of contact during the often lengthy process. A hotline may even be established if significant concerns or questions on the part of the community are anticipated.

Proposed Remedy Selection The agency selects and proposes a remedy at the conclusion of the corrective measures study. The agency must then initiate a modification of the facility's permit to incorporate the proposed remedy. This modification requires the same level of public involvement as is required for a draft permit. The permit with the proposed modification must be released for public review and comment, and the community must receive public notice of its availability through a major local newspaper and through broadcast by local media. A fact sheet or statement of basis must be prepared to explain the proposed permit modification and the significant factual, methodological, policy, and legal reasons for the remedy chosen.

A 45-day comment period on the draft permit modification follows publication of the notice to provide the community with an opportunity to comment, in writing, on the remedy described. The comment period may be extended or reopened if significant questions or concerns are raised. A public hearing may be requested and, if so, a 30-day advance notice is required.

The agency also recommends holding workshops or small group meetings to explain the proposed remedy to the community. Such a proactive approach can significantly enhance the decision-making process and can reduce the possibility of opposition to the remedy ultimately selected. This is especially important when the remedy proposed is controversial or when public concern about health and safety issues in conjunction with the facility is high.

Final Remedy Selection The final remedy is selected at the conclusion of the public comment period and after review of all the comments received. The agency must send a notice of decision to the facility owner or operator and to any persons who submitted comments or requested notification. The agency also prepares a written response to comments that summarizes and explains how the comments were either addressed or rejected in the final permit modification for the remedy selected.

Implementation of the Corrective Measures or Remedy. This phase proceeds after the permit modification is approved. No additional public involvement activities specifically related to implementation of the remedy are required. Depending on the remedy chosen, however, additional public involvement activities may be deemed necessary.

Remediation of significant contamination at a facility may take years to complete. As such, it makes sense to involve the public in the steps along the process, or at least at the significant milestones. Fact sheets are often used to keep the public informed about the pace of cleanup efforts and to alert the community when significant events, such as the startup of a groundwater pump-and-treat system, occur. Availability sessions can be used to provide the community with the opportunity to ask questions and express concerns to experts and other parties involved in the remediation activities. Additionally, depending on the remedial action, the facility may host an open house to allow the public to view parts of the remediation setup. Clearly, safety concerns must be considered before members of the community are invited on-site.

The proposed Subpart S rule requires that all individuals on the mailing list be notified that the construction plans and remedy specifications are available for public inspection. This information must also be placed in the chosen repository.

Upon completion of the remedial activities, the owner or operator must request another permit modification in order to remove the compliance schedule and other pertinent information related to the remedial or corrective action activities from the permit. This type of modification is considered to be a Class 3 modification and subject to the same public involvement activities.

Public Involvement Under Enforcement Orders The agency issues what is termed a Section 3008(h) enforcement order to require corrective actions at nonpermitted (interim status) RCRA facilities where there is evidence of a hazardous waste or hazardous constituent release. This order can be issued either on a consent basis or unilaterally. A consent order is issued when the facility and the agency agree on the necessary corrective action, while a unilateral order is issued when the agency and facility are unable to agree about the scope of the necessary corrective actions.

No public involvement activities are specifically required under Section 3008(h), although they are strongly encouraged. Public involvement activities typically utilized during enforcement actions are similar to those used for corrective actions. However, there may be legal limitations on the type and amount of information that can be released to the public under an enforcement order. A facility can agree to conduct public involvement activities under a consent order, but under a unilateral order, the agency typically has responsibility for such activities.

Equitable Public Participation and Environmental Justice

The agency has chosen to address equitable public participation and environmental justice issues related to RCRA permitting provisions in guidances rather than through regulation. Current guidances and documents recommend the use of multilingual notices and fact sheets, along with the use of translators for meetings, in areas that contain significant numbers of affected community members who do not speak English as a first language. The newer requirement for an informal public meeting prior to permit application is an attempt to focus efforts on expanding equitable public participation and ensuring that environmental justice issues (if any) are confronted as early in the process as possible.

The agency maintains that the best way to approach environmental justice issues is at the local level and on a site-specific basis, rather than through regulation. State and regional EPA offices have been charged with establishing mechanisms to respond effectively to environmental justice concerns during permitting activities. RCRA implementing agencies continue to address patterns of disproportionately high and adverse environmental effects and human health impacts on low-income communities as a result of hazardous waste management activities. State and regional offices have also been involved in environmental justice projects that emphasize reaching out to affected communities with public involvement activities. Environmental justice is discussed in greater detail in chapter 8.

REQUIREMENTS UNDER CERCLA/SARA AND THE NCP

Community relations requirements and policies are set forth in the Comprehensive Environmental Response, Compensation, and Liability Act (CERCLA), which was originally promulgated in 1980 (and updated through the Superfund Amendments and Reauthorization Act (SARA) in 1986) and also promulgated in the National Oil and Hazardous Substances Pollution Contingency Plan (NCP), which was updated most recently in September 1994. Superfund is a term that is often used interchangeably with CERCLA and SARA. The applicable community relations provisions are primarily contained in 40 CFR Part 300, and related EPA policies are also provided in several guidances and manuals. The following discussion presents the essential components of the community relations activities that are required within these regulations and guidances. First, a brief overview of the major regulatory provisions is provided.

Regulatory Provisions

Community relations activities under CERCLA/SARA and in the NCP relate, essentially, to two general categories of actions. The first corresponds to activities related to actions taken as part of a *remedial response* under Superfund. Remedial responses basically encompass actions that are part of the process of identifying, investigating, and remediating sites on the National Priorities List (NPL) for Superfund. The NPL is the EPA's list of the most serious uncontrolled or abandoned hazardous waste sites identified for possible long-term remedial response. Listing on the NPL is based on the score that the site receives using a standardized Hazard Ranking System (HRS) analytical process.

A remedial response at a Superfund site can be a costly and time-consuming process that may take years to reach completion. There are a number of "milestones" in the remedial response process, each with unique characteristics that require specific community relations activities. These milestones are:

- Site assessment activities (includes preliminary assessment through initial investigation and HRS scoring to include the site on the NPL)
- Remedial investigation of the site

- Feasibility study for potential remedial actions
- Remedial design, remedial action, and operation and maintenance of remediation system
- Completion of remediation and removal of site from NPL

At each milestone, different kinds of issues and concerns may arise; thus, certain community relations activities are required and/or recommended during each phase.

Remedial responses may also be undertaken through a CERCLA *enforcement action*. The EPA has the authority to identify potentially responsible parties (PRPs) who may have owned or operated a hazardous waste site or generated, transported, or disposed of hazardous substances. CERCLA grants the EPA the authority to negotiate settlements for site cleanup work or to issue *administrative orders* directing the PRPs to participate in the remedial process, either through performance of the actual work or through payment of remediation costs. Sites undergoing remediation under an enforcement action sometimes require additional community relations efforts, as there may be a significant degree of friction between the community and the PRPs.

The second category of activities corresponds to that associated with *removal actions*. Removal actions are taken as emergency or short-term responses to an immediate threat to public health, welfare, and the environment. Removal actions can occur at a Superfund site, in response to a spill or accident that involves hazardous waste, and as a result of other scenarios where the unexpected discovery of hazardous wastes is considered to present an immediate threat. Removal actions are different from remedial actions in that they are typically limited in both scope and cost. The action is intended to secure the removal of the hazardous waste in as short a time frame as possible, in order to protect the health and welfare of the public and environment in what is determined to be an "emergency" situation.

The NCP establishes the general requirements for all removal actions, and specific requirements for three types of removal actions:

1. Removals with a planning period of less than six months.
2. Removals expected to extend beyond 120 days.
3. Removals with a planning period of at least six months.

The following discussion presents the community relations activities that are either required or recommended during the different phases of a remedial response and during removal actions.

Community Relations During Site Assessment Activities

Community relations activities undertaken before the commencement of the remedial investigation phase of the remedial response process depend, in large part, on-site specific circumstances. There are instances in which public interest is high from the beginning of the process—sometimes, community involvement is the catalyst that first brings the site to the attention of the regulatory agency. In such a scenario, the agency may engage in community relations activities early in the site assessment and investigatory phases. Figure 7.3 displays the community relations activities required during each stage of a remedial response.

Preliminary Assessments   The first step in the site assessment process is to conduct a preliminary assessment (PA). The PA involves a review of site records and other existing permits or analytical data in order to evaluate past activities at the site and determine whether further investigation is required. No on-site visit or sampling is usually conducted at this stage. Formal community relations activities are not typically conducted during this phase; however, the agency indicates that certain sites may require early efforts based on the following factors:

- The likelihood that the site will be included on the NPL
- The site's location with respect to other NPL sites
- The site's location with respect to certain population centers
- The amount of current community interest in, or concern about, the site

Sometimes, the agency may make preliminary contacts with local officials and community leaders to inform them about the regulatory process and to gather information about community concerns related to the site.

Site Inspection   The site inspection is performed at the conclusion of the PA activities at sites where information gathered indicates that there could be a potential hazard. Site inspection activities are undertaken in order to obtain additional information for use in scoring the site for possible inclusion on the NPL and/or to determine if a removal action is necessary.

The site inspection typically involves a team of investigators who may spend several days at the site conducting sampling and other field activities.

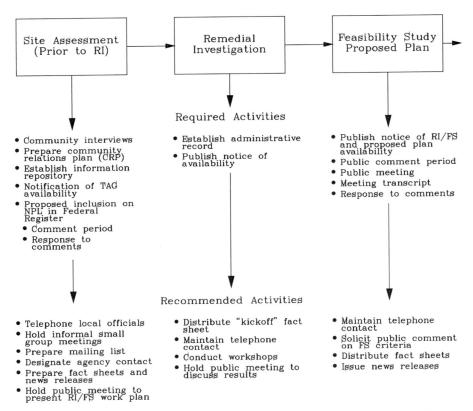

■ FIGURE 7.3 Community Relations Activities During CERCLA

Such activities can be highly visible to surrounding neighbors and other community members. At this stage, the agency may begin to conduct interviews with local officials, community residents, public interest groups, and other interested or affected parties. The purpose of the interviews is to assess the level of community concern and determine how and when the community wants to be involved in the Superfund process.

A fact sheet may be designed to explain the purpose of the site investigation activities and possible outcomes. In this way, the agency continues to lay the groundwork for upcoming phases of the remedial response. Additional activities that can occur during the site investigation stage include setting up a hotline so community members can report information or ask questions; creating a mailing list; and designating a contact person to respond to community concerns.

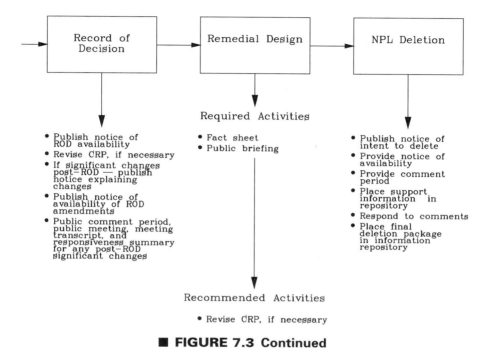

■ FIGURE 7.3 Continued

Often, a significant time lag can occur between the initial site investigation activities and the final report and HRS score. During this time, the agency may keep interested and affected persons updated on the progress of the investigation through additional fact sheets and phone contact.

A local information repository needs to be established before the start of the remedial investigation activities, and may be created during the site investigation stage. The information repository must be established at a location near the site and be accessible to local community members and affected parties. The repository should contain a copy of all available information, including reports, data, and fact sheets. The agency is responsible for informing interested parties about the presence and location of the information repository.

NPL Listing Process The agency proposes placement of a site on the NPL through publication of the proposal in the Federal Register. A public comment period of 60 days is required in order to provide the community with the opportunity to respond to the proposed inclusion. The agency must respond to all written public comments through a responsiveness summary that addresses all significant comments and indicates how they affected, or did not affect, the final decision. The agency must make any revisions resulting from the comments received and publish the final rule in the Federal Register no less than 30 days prior to the effective date of the site listing.

At the time of the proposed listing, the agency may decide to develop an additional fact sheet to explain the listing process and how the site was chosen for inclusion. The fact sheet should also explain the comment process. Depending on the expressed level of interest, the agency may also choose to conduct a formal or informal public meeting and prepare a news release for the media.

Community Relations During Remedial Investigation and Feasibility Study Activities

The purpose of the remedial investigation (RI) is to collect additional data necessary to characterize the site in order to develop and evaluate remedial alternatives. The RI may be conducted in one or more phases in order to provide the information necessary to assess the risks to human health and the environment and to support the development, evaluation, and selection of appropriate remedial alternatives.

A number of community relations activities must be completed before beginning remedial investigation site work. A mailing list must be compiled, community interviews must be completed, and the information repository must be established at some point during the assessment phase and prior to the remedial investigation. The results of the community interviews are used to produce a formal community relations plan (CRP). The CRP specifies the community relations activities that the agency expects to undertake during the upcoming remedial investigation. The purpose of the CRP is to:

- Ensure that the public has an opportunity to be involved in the process of site characterization, analysis, and remedial selection activities;
- Determine the most appropriate activities to ensure public involvement and obtain community support; and
- Provide opportunities for the community to learn about the regulatory process and the site under investigation.

The agency must also inform the public about the availability of technical assistance grants (TAGs) and include material in the information repository that describes the TAG application process. The TAG program provides funds for qualified citizens' groups to hire independent technical advisors to help them understand and comment on technical decisions related to Superfund actions.

Before the remedial investigation begins, a draft work plan that explains all planned activities must be prepared, usually by a government contractor, and approved by the agency. When the draft work plan and the CRP are completed, the agency usually holds a public meeting to explain the plans to the community. The agency may choose to distribute a "kickoff" fact sheet and use other communication methods, depending on the level of community interest. Additional fact sheets are usually developed and sent to all interested or affected parties as the investigation progresses to completion. Small group meetings and workshops may also be held to educate the community about the investigative process if interest about the site is high.

The agency must establish an *administrative record* and maintain it in the information repository. The administrative record is a file that contains all of the information used by the agency in making its site-related decisions. The file must be available for public review, and the agency must publish a notice of its availability in a major local newspaper of general circulation.

The remedial investigation can take a considerable amount of time to complete and delays in data receipt and reporting are common. During lag times, the agency usually attempts to keep the public interested and involved by maintaining communication through the distribution of fact sheets and conducting small group meetings. Upon completion of the remedial investigation report, a public meeting is held to explain the results to the community.

A feasibility study (FS) is undertaken at the conclusion of the remedial investigation activities for the purpose of identifying possible remedial alternatives. The primary objective of the FS is to ensure that information concerning the potential remedial options is thoroughly developed and evaluated so that decision makers can select the most appropriate remedy. The FS may address the site in its entirety or focus on specific areas.

Public interest typically increases during this stage of the process, when actual remedial options are being considered. Additionally, the potential for conflict increases as the development of different cleanup alternatives (each with positive and negative attributes) continues. Suggested community relations activities during this period include:

■ Conducting informal public meetings with community members, local officials, and other affected parties

- Issuing news releases, fact sheets, and publicly available progress reports
- Maintaining and updating materials in the information repository and administrative record
- Holding availability sessions to respond to individual questions and concerns

Availability sessions can provide an important opportunity for community members to meet with remediation experts to discuss the pros and cons of potential remedial options. Such an opportunity can significantly raise the community's level of comfort with the remediation option ultimately chosen.

Community Relations upon Completion of RI/FS

Upon completion, the remedial investigation and feasibility study reports are placed in the information repository for public review. The first step in the actual remedy selection process is the development of a *proposed plan* that summarizes the remedial alternatives presented in the detailed analysis of the feasibility study, identifies the preferred alternative, provides the rationale for choosing that alternative, identifies any proposed waivers to cleanup standards, and documents agency comments. The agency must accomplish the following community relations activities upon completion of the proposed plan:

- Publish a notice of the availability of the proposed plan and RI/FS reports, a brief summary of the proposed plan, and an announcement of a 30-day comment period. The notice must be published in a major local newspaper of general circulation.
- Make the proposed plan and supporting analysis and documentation available in the administrative record and information repository.
- Provide the opportunity for submission of both oral and written comments—a formal period of not less than 30 days is required and the agency must extend the comment period by at least 30 days upon a timely request.
- Provide the opportunity for a public meeting to be held during the public comment period—the agency must prepare a transcript of all formal public meetings held during the comment period and make it available to the public.

■ Prepare a responsiveness summary that incorporates significant public comments and the agency's response to those comments that becomes a permanent part of the record of decision.

The record of decision (ROD) is a public document that explains which remedial alternative has been selected. The ROD is based on information and technical analysis generated during the RI/FS and upon consideration of public comments and community concerns.

The agency must make the ROD available for public inspection and review at the information repository. The agency must also publish a notice of the ROD's availability in a major local newspaper of general circulation. The notice must state the basis under which the remedial alternative was selected. Prior to the remedial design phase, the agency must revise the community relations plan, if necessary, to address concerns particular to the design and construction phase of the selected remedial action.

Significant Changes Prior to ROD If new information obtained during the public comment period significantly changes the basic features of the remedy as originally presented in the RI/FS and proposed plan, the agency is required to document those changes in the ROD. The regulations require an additional public comment period if those changes are not "reasonable extensions" of the information originally presented in the RI/FS and proposed plan. The agency must also issue a revised proposed plan which includes a discussion of, and reasons for, the significant changes.

Significant Changes Post-ROD When the remedial or enforcement action, or the settlement or consent decree, differs significantly from the remedy selected in the ROD with respect to scope, performance, or cost, the agency must publish a notice that briefly summarizes and explains the significant differences and the reasons for the change. The notice must be published in a major local newspaper, and supporting information and documentation must be made available to the public in the information repository and administrative record.

When the remedial or enforcement action, or the settlement or consent decree, fundamentally alters the basic features of the selected remedy with respect to scope, performance, or cost, the agency must propose an amendment to the ROD. The agency must also issue a notice of availability and a brief description of the proposed amendment in a major local newspaper of general circulation. The same procedures as required for completion of the RI/FS report and proposed plan must be followed with respect to the public comment period, public meeting, transcript of comments, and responsiveness summary.

The agency must publish a notice of the amended ROD's availability at the conclusion of the comment period and presentation of the responsiveness summary. The notice must be published in a major local newspaper and the amended ROD and supporting information must be made available for public inspection in the information repository and administrative record.

Community Relations During Remedial Design, Remedial Action, and O&M Activities

During the remedial design phase, responsibility for the site typically shifts from the federal or state lead agency to the U.S. Army Corps of Engineers, U.S. Bureau of Reclamation, or other remedial design contractor. This kind of major change of participants may necessitate additional community relations activities to ensure that community questions and concerns continue to be addressed.

Prior to the completion of the remedial design, the agency may revise the CRP to address new or changing community concerns more specific to the actual remediation of the site. If concerns are essentially unchanged, the revision may be limited to providing new or updated activities and schedules based on the selection of the remedial option. If selection of the remedial option has generated controversy or resulted in a significant change in community attitudes, however, a more substantial revision may be necessary. Additional community interviews may be conducted to provide input into the necessary CRP changes.

When the final engineering design is complete, the agency issues a fact sheet that provides a description of the final remedial design. The agency also provides a public briefing, as appropriate, before beginning the remedial action activities in order to describe the remedial design to the community.

There are no formal community relations activity requirements during the remedial action. It can take years to complete a site cleanup, however, and activities are usually continued to keep the community interested and informed about the progress of remediation. Effective tools that are commonly used include periodic updates at significant milestones in the form of fact sheets and news releases; informal public and small group meetings; and availability sessions to allow individuals to ask questions and discuss concerns. Other suggested activities involve holding open houses or conducting site tours to allow members of the community to see for themselves how the remedial action is progressing. A pictorial history that displays the various stages of the process can effectively demonstrate the progress that has been made over the years.

The public can be made aware of operation and maintenance (O&M) issues associated with the ongoing remedial action through periodic updates and published schedules of O&M activities. Fact sheets can be used to explain the shutdown and startup procedures of the remediation technology. Small group meetings may be held periodically to enhance public understanding of the processes.

Community Relations During the NPL Deletion Process

Procedures for deleting a site from the NPL are similar to those required to make an addition to the NPL. The agency must prepare public notification statements, and a *notice of intent to delete* must be published in the Federal Register and in a local newspaper of general circulation. The agency must also provide at least a 30-day comment period on the proposed deletion. Copies of information supporting the proposed deletion must be placed in the information repository for public review. The agency must respond to significant comments and any new data submitted during the comment period and include the responsiveness summary in the final deletion package. The responsiveness summary should describe:

- Comments received during the public comment periods and public meetings;
- Agency responses to national and local public comments; and
- Justification for proceeding with the deletion—particularly if public comments indicate disagreement with the recommendation to delete.

Following approval by the EPA regional administrator, a copy of the responsiveness summary is made available in the regional docket and published as part of the final rule. The final deletion package must be placed in the local information repository once the notice of final decision has been published in the Federal Register.

Community Relations During Enforcement Actions

Remedial responses may also be undertaken through a CERCLA *enforcement action*. The EPA has the authority to identify potentially responsible parties (PRPs)—individuals or companies (such as owners, operators, transporters, or generators of hazardous waste) potentially responsible for, or contributing to, the contamination of a Superfund site. Whenever possible, the EPA requires PRPs, through administrative or enforcement actions, to clean up hazardous waste sites for which they are responsible.

CERCLA grants the EPA the authority to negotiate settlements for site cleanup work or to issue *administrative orders* directing the PRPs to engage in site remediation work or pay for associated costs. The agency in charge of the response action develops and conducts community relations activities at enforcement-lead sites. PRPs may participate in community relations activities only at the discretion of the regional office, and the agency must develop the community relations plan. The PRPs may wish to participate in public meetings and/or in the preparation of fact sheets; however, the agency must review all written material before it is released to the public.

The completed CRP should reflect the PRPs' involvement and must be provided to all interested parties and made available in the administrative record and information repository (along with any subsequent revisions).

The public must be informed when agreements or settlements are reached between the agency and PRPs. Notice is sent to all parties on the mailing list, and a press release may be issued. RI/FS settlements generally take the form of an *administrative order on consent* (AOC). This is a legal agreement between the EPA and PRPs, in which the PRPs agree to perform and/or pay the cost of the site cleanup. The agreement describes the actions to be taken at the site and may be subject to a public comment period.

When the agency reaches a settlement with the PRP to perform RI/FS activities, a "kickoff" meeting with the public is held to explain the AOC and outline the remedial process. Special issues that usually need to be clarified include the EPA's approval of the PRPs' work plan, the responsibilities of the PRPs in the performance of the RI/FS, and agency oversight of the work. Other community relations activities proceed as described previously with regard to the remedial response process and as appropriate to site-specific circumstances.

The ROD is published upon completion of the RI/FS and proposed plan. After publication, the agency attempts to reach an agreement with the PRPs with respect to implementing the remedial design and remedial actions selected. When a negotiated settlement is reached, a *consent decree* is submitted to the U.S. District Court for approval. A consent decree is a legally binding agreement between the PRPs and the agency.

The delay between the time the consent decree is referred to the Department of Justice (DOJ) and lodged with the court may extend to several months. A press release may be issued in the interim to inform the public about the settlement terms. At the time the DOJ lodges the consent decree with the court, a notice of the proposed agreement must be published in the Federal Register. A public comment period is required before the proposed consent decree is entered by the court as a final judgment. The public comment period must be at least 30 calendar days in length, and may be extended upon request. The proposed consent decree may be withdrawn or modified if it is determined to be inappropriate or inadequate, based on comments received.

Copies of the proposed consent decree are distributed to persons on the mailing list. Procedures for public comment on the consent decree and a contact name are included. The agency may elect to hold a public meeting during the comment period if significant community questions or concerns arise.

Once the public comment period on the proposed consent decree closes, the DOJ considers each significant comment and prepares a response. The DOJ then files a *motion to enter* the consent decree, a response to comments, and summary of significant comments received. The motion to enter and response to comments are released to the public at the same time. All documents must be made available to the public in the information repository.

Community Relations During Removal Actions

Removal actions comprise a separate category of response that is initiated on an emergency or short-term basis in order to act on an immediate threat to public health and welfare, or to the environment. Removal actions are usually limited in time to no more than one year and in scope to removal of the threat. As discussed, the NCP establishes general requirements for all types of removal actions and more specific requirements for three categories of action:

1. Removals with a planning period of less than six months.
2. Removals expected to extend beyond 120 days.
3. Removals with a planning period of at least six months.

The categories comprise distinct types of removal actions, which range from an emergency response that must be undertaken within several hours (such as that associated with a tanker truck spill) to a situation involving a less critical response time (such as draining a contaminated lagoon or pond).

When the EPA becomes aware of an emergency situation, a preliminary assessment (PA) is conducted to evaluate the site in order to determine the nature and extent of the hazard. The PA is used to determine whether federal action is necessary, and usually involves reviewing site management practices and gathering other available information on site operations and activities. The removal PA may include identification of the source and nature of the threat or release; evaluation of the magnitude of the threat; evaluation of factors necessary to determine whether removal is necessary; and a determination of whether a nonfederal party is undertaking an appropriate response.

A removal site inspection may be performed if more information is required. This inspection can involve an off-site reconnaissance or on-site investigation, depending on safety considerations. The agency's lead on-scene coordinator (OSC) evaluates the information and determines whether a federal removal action is necessary to prevent or mitigate the threat.

Different kinds of community relations activities are required, based on the expected duration of the removal action. The activities are intended to ensure that the community is kept informed about the process. Figure 7.4 provides a synopsis of community relations requirements for removal actions.

Requirements for All Removal Actions In all removal actions, the agency must designate a spokesperson to inform the community about the

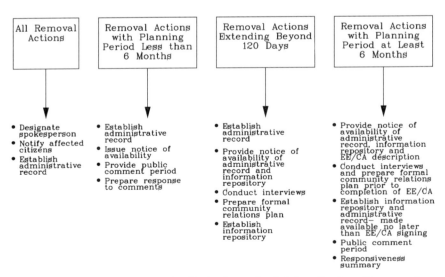

■ **FIGURE 7.4 Community Relations Activities During Removal Actions.**

actions taken, respond to inquiries, and provide information concerning the hazardous substance release. Notification of this information must be made to affected community members, state and local officials, and, if appropriate, civil defense or emergency management agencies. Any news release or statements made by the participating agencies must be coordinated with the OSC or regional project manager (RPM). The agency also needs to establish an administrative record and make it available to the public at a location that is convenient and accessible. Notification of the availability of the administrative record must be published in a major local newspaper of general circulation. The community must also be informed if an information repository is created.

Removal Actions with a Planning Period of Less than Six Months

When a removal action begins within a six-month period, the agency must notify the public about the availability of the administrative record, provide a public comment period, and prepare a responsiveness summary of comments received.

The agency must publish a notice of availability of the administrative record in a major local newspaper of general circulation within 60 days of initiating on-site removal activity. The administrative record is made available for public inspection and review at a central location. The agency must provide a public comment period of at least 30 days, if appropriate, from the time the administrative record is made available for review. The agency then prepares responses to significant comments and any new information submitted during the public comment period. The summary of responses is placed in the administrative record for public review.

Removal Actions Expected to Extend Beyond 120 Days

Additional community relations activities are required at sites where the removal actions are expected to extend beyond 120 days. By the end of the first 120 days of the action, the agency must conduct interviews with local officials, community residents, public interest groups, and other interested or affected parties in order to identify the community's specific concerns and information needs. The agency uses this information to develop a formal community relations plan. The CRP specifies the community relations activities that the agency expects to undertake during the removal action. The CRP must also be completed within 120 days of the start of on-site activities.

The agency is also required to establish an information repository at or near the location of the response action. The repository contains information available for public review in addition to the administrative record. The agency must inform the public that it has established the information repository and provide notice of the availability of the administrative record file for review and copying.

Removal Actions with a Planning Period of at Least Six Months
Additional community relations activities are required at sites where at least six months will elapse before the start of the planned removal actions. The same CRP requirements as specified for actions beyond 120 days apply, with the exception that the interviews must be conducted and the plan finished prior to completion of the engineering evaluation/cost analysis (EE/CA) report that is prepared to analyze the removal alternatives. Similarly, the agency must also establish the information repository and administrative record, with the exception that the activity must be completed no later than the time that the EE/CA approval memorandum is signed.

The agency is required to issue a notice of availability and a brief description of the EE/CA in a major local newspaper of general circulation. A comment period of at least 30 days must be provided for submission of written and oral comments concerning the EE/CA and planned site activities. The agency may extend the comment period by a minimum of 15 days if a "timely request"—one made within two weeks of the end of the comment period—is made.

The agency prepares a written response to significant oral and written comments made during the public comment period, and must make the document available to the community in the information repository.

NPDES PROGRAM REQUIREMENTS

The public involvement requirements for programs administered under the Clean Water Act are contained in 40 CFR Part 124. The following discussion provides the general program and public involvement requirements associated with the National Pollutant Discharge Elimination System (NPDES) permit program.

General Requirements
Under the NPDES program, permits must be issued to any facility that discharges pollutants from any point source into the waters of the United States. The permit program also applies to owners or operators of any treat-

treatment works responsible for treating domestic sewage, whether or not the treatment works is required to obtain a permit under authorization of other provisions.

Public Involvement Requirements

When a request for an NPDES permit is submitted, the agency must decide to either prepare or deny the draft permit. If the decision is to deny the permit, a notice of intent to deny must be issued.

The agency must provide a public notice of the permit action and public comment period. Public notice is required when the permit application has been tentatively denied; a permit has been prepared in draft; a hearing has been scheduled; an appeal has been granted; or an NPDES new source determination has been made. No public notice is required when a request for permit modification, revocation and reissuance, or termination is denied. In such cases, written notice of the denial is provided only to the permittee and anyone who specifically requests it.

The provision of public notice about the preparation of the draft permit must allow at least 30 days for public comment and be made at least 30 days before a scheduled public hearing. Public notice of the hearing may be combined with the notification of the draft permit and comment period. The notice must be mailed to the applicant, any applicable public agency, any user of a privately owned treatment works identified in the permit application, and persons on the mailing list. The mailing list is typically composed of persons who request to be included, participants in past permit proceedings in the applicable area, and local elected officials. The notification must also be published in a major local newspaper of general circulation.

If the agency decides to prepare the draft permit, a fact sheet must be prepared for distribution to affected parties. The fact sheet must include, as applicable, a brief description of the facility and operation that is the subject of the draft permit, along with the type and quantity of waste, fluids, or pollutants that are proposed for discharge. It needs to provide a brief summary of the basis for the draft permit conditions, including references to applicable regulatory provisions and supporting documentation. The fact sheet also describes reasons why any requested variances or alternatives to required standards are justified. A description of the procedures for reaching a final decision on the draft permit must be provided and includes:

1. The beginning and ending dates for the public comment period and the address to which comments should be sent.

2. Procedures for requesting a hearing and other mechanisms for public participation in the final decision-making process.

The fact sheet must also contain the name and telephone number of a person to contact for additional information.

Any calculations or other explanations for the derivation of effluent limitations and conditions or standards for sewage sludge use or disposal must be included. A citation of the applicable effluent limitation guideline or standard should be made, along with the reasons for its applicability or an explanation of how alternate effluent limitations were developed.

The fact sheet must provide an appropriate explanation whenever the draft permit contains any of the following conditions:

- Limitations to control toxic pollutants
- Limitations on internal waste streams
- Limitations on indicator pollutants
- Limitations set on a case-by-case basis

A sketch or detailed description of the discharge location or regulated activity, as specified in the permit, should also be provided in the fact sheet. For permits that include a sewage sludge land application, a brief description of how each of the required elements of the land application plan is addressed in the permit should be provided.

During the public comment period, any person can submit written comments on the draft permit and request a public meeting. A public meeting request must be made in writing. The agency may independently decide to conduct a public meeting if community questions or concerns about the permit conditions are anticipated. The public comment period is automatically extended to the conclusion of any scheduled public meeting. The agency may also decide to reopen or extend the comment period based on the level of community interest. A tape recording or written transcript of the meeting must be made available to the public.

After the close of the comment period, the agency must issue a final permit decision. The permit applicant must be notified in writing, as must each person who submitted written comments or requested notification of the decision. The agency must also issue a response to all significant comments. The response must specify which provisions of the draft permit have been changed as a result of the comments, and explain why other comments were not incorporated.

The final permit and response to comments must be made available in the administrative record, along with supporting documentation and meeting transcripts.

EMERGENCY PLANNING AND COMMUNITY RIGHT-TO-KNOW ACT

Legislation for the Emergency Planning and Community Right-to-Know Act (EPCRA), also known as Title III of the Superfund Amendments and Reauthorization Act (SARA), was passed by Congress in 1986. The act establishes provisions for federal, state, and local governments and industry with respect to emergency planning and reporting requirements concerning hazardous and toxic chemicals. The provisions are intended to increase the public's knowledge about, and access to information on, the presence and use of hazardous chemicals in their communities and the potential for release of these chemicals into the environment. A brief overview of the four major EPCRA provisions is presented, followed by a discussion of the public involvement and reporting requirements within the community right-to-know program.

EPCRA Provisions

The EPCRA provisions are contained under four primary headings:

1. Emergency planning
2. Emergency release notification
3. Community right-to-know reporting requirements
4. Toxic chemical release inventory reporting

The EPCRA provisions build on the EPA's Chemical Emergency Preparedness Program (CEPP) and numerous state and local programs designed to help communities understand and prepare for potential chemical emergencies. Facilities that meet certain requirements, such as using or handling listed chemicals above predetermined threshold quantities, are subject to the regulations.

Emergency Planning Provisions The *emergency planning provisions* are contained in Sections 301-303 of SARA and promulgated in 40 CFR

Part 355. These provisions are designed to enhance the capabilities of state and local governments and the community in responding to emergencies involving hazardous substances. The governor of each state appoints members to a state emergency response commission (SERC), which is charged with the oversight of state emergency planning activities. Representatives of existing state agencies, such as environmental, health, or transportation, may be named to the SERC. Members of trade associations, public interest organizations, and other public and private sector groups with emergency planning knowledge or experience can also be included on the SERC.

In turn, each SERC divides its state into local emergency planning districts and designates local emergency planning committees (LEPCs) for each district. The LEPC membership must include, at a minimum:

■ Elected state and local officials or representatives
■ Law enforcement officials, civil defense workers, and firefighters
■ Public health, hospital, transportation, and environmental professionals
■ Representatives of community groups and media
■ Owners, operators, or other representatives of facilities subject to the emergency planning requirements

The LEPCs are charged with organizing chemical data collected from the affected facilities and developing emergency response plans for their communities. These plans are known as local emergency response plans (LERPs) and present the potential hazards, response capabilities, and procedures that the affected community and facilities need to follow in a hazardous chemical or substance emergency. Each plan must:

■ Identify hazardous substance facilities and transportation routes
■ Describe on-site and off-site emergency response procedures, notification protocol, and evacuation plans
■ Designate a community coordinator and affected facility coordinator(s) to implement the plan
■ Describe methods for notification about the release and probable affected areas and populations, as well as the emergency equipment and response personnel available
■ Provide a training program and schedule for emergency response personnel, in addition to presenting methods and schedules for exercising the plan

Emergency Notification Provisions The *emergency notification provisions* are contained in Section 304 of SARA and promulgated in 40 CFR Part 355. Under these provisions, facilities must immediately notify the LEPCs and SERCs if there is a release into the environment of a regulated hazardous substance that exceeds a "reportable quantity." Substances subject to this requirement are those on the list of extremely hazardous substances published in 40 CFR Part 355, as well as those that are subject to the emergency notification requirements under CERCLA Section 103(a)—found in 40 CFR Part 302. Some chemicals are common to both lists. The CERCLA hazardous substance regulations also require that release notifications be made to the National Response Center (NRC), which in turn alerts federal responders.

Requirements for some of the most hazardous and toxic chemical substances found on these lists provide that releases of more than one pound be reported. Other chemicals have reportable quantities that range from 10 to 10,000 pounds. The notification of a release activates the emergency response plans that are prepared by the LEPCs.

Community Right-to-Know Reporting Requirements EPCRA contains two separate but related *community right-to-know reporting requirements*. Section 311 of SARA, promulgated in 40 CFR Part 370, requires that the material safety data sheets (MSDSs) prepared by facilities under the Occupational Safety and Health Administration's (OSHA's) Hazard Communication Standard (HCS) regulations be submitted to the LEPC, the SERC, and the local fire department. The MSDSs contain information on a chemical's physical properties and health effects.

In lieu of the actual MSDSs, facilities have the option of providing lists of chemicals for which MSDSs have been prepared. However, the actual MSDS must be presented, upon request, by the LEPC, SERC, or local fire department. The MSDSs or list must be submitted for any chemical that is present at the facility above certain threshold levels. The EPA has established a threshold of 500 pounds (or the threshold planning quantity, whichever is lower) for extremely hazardous substances and 10,000 pounds for all other hazardous chemicals.

Under the HCS, chemical manufacturers and importers must research the chemicals they produce and import. If a substance presents any physical or health hazard as specified in the HCS, then information about that hazard must be communicated to employees, as well as subsequent purchasers or users of the chemicals. OSHA requires companies to keep MSDSs on file for all hazardous chemicals in the workplace. MSDSs are commonly found in many types of workplaces and must be made available to all employees for their health and safety protection.

Section 312 of SARA, promulgated in 40 CFR Part 370, provides the second type of reporting requirement. Facilities must submit an emergency and hazardous chemical inventory form to the LEPC, the SERC, and the local fire department on an annual basis. Hazardous chemicals covered by the regulation are those for which the facility is required to prepare MSDSs under OSHA's Hazard Communication Standard. The chemicals must also have been present at the facility at some time during the previous year at a level above the thresholds described previously.

These regulations take a "two-tier" approach for annual inventory reporting. Under Tier I, a facility must report the amounts and locations of chemicals in certain hazard categories. The Tier II report contains basically the same information, but names the specific chemicals. Companies have the option of choosing to file either Tier I or Tier II forms. However, the owner or operator of a facility must submit a Tier II form to the SERC, LEPC, or local fire department upon request.

Toxic Chemical Release Reporting Provisions The *toxic chemical release reporting* requirements are contained in Section 313 of SARA and promulgated in 40 CFR Part 372. Section 313 requires the EPA to establish an inventory of routine toxic chemical emissions—termed the Toxics Release Inventory (TRI). The Pollution Prevention Act of 1990 significantly expanded the TRI by requiring additional reporting in an effort to differentiate among source reduction, recycling, treatment, and disposal options. Mandatory information on source reduction, recycling, and treatment methods must now be collected and reported on the TRI form. For reporting purposes, a release is defined as any spilling, leaking, pumping, pouring, emitting, emptying, discharging, injecting, escaping, leaching, dumping or disposal into the environment. This includes the abandonment or discarding of barrels, containers, and other closed receptacles of any listed chemical covered by the regulations.

The reporting requirement applies to owners and operators of facilities that have 10 or more full-time employees, and that manufacture, process, or use more than specified threshold quantities of more than 600 listed chemicals. Facilities importing, manufacturing or processing any of these chemicals in excess of 25,000 pounds are required to submit the form. Facilities "otherwise using" the listed toxic chemicals in quantities above 10,000 pounds are also required to submit the release form annually.

Facilities subject to this reporting requirement must complete a Toxic Chemical Release Inventory Form (Form R). The following information is required on Form R:

- Name and location of facility and identity of the listed chemical (unless claimed as a trade secret)
- The maximum quantity of the chemical on-site at any time during the year and total quantity of chemical released during the year, including both accidental spills and routine emissions
- Off-site locations where wastes containing the chemical were shipped and the quantities sent to those locations for recycling, energy recovery, treatment, or disposal
- On-site recycling, energy recovery, treatment, or disposal methods used for wastes containing the listed chemical and estimates of treatment efficiency for each chemical
- Quantities of each chemical recycled, combusted for off-site energy recovery, treated, and released on-site and off-site
- Source reduction activities involving the chemical

Facilities must estimate the total amount of listed chemicals that they release into the environment each year. That quantity can be as a result of routine operations, accidental release, or through transport of waste to off-site locations. The purpose of the TRI reporting requirement is keep the public and government officials informed about releases of toxic chemicals to the environment.

Public Involvement for Emergency Planning and Notification

The LEPC for a given community is required to complete tasks in addition to preparing an emergency response plan. Those tasks include providing public notice of its activities and establishing procedures for handling public requests for information.

The emergency response plan must initially be reviewed by the SERC and be updated at least annually by the LEPC. The LEPC must publicize the plan through public meetings or newspaper announcements, receive and consider public comments, and periodically test the plan by conducting emergency drills. Community members can volunteer to serve on the LEPC as citizen representatives and have the right to review and comment on the emergency response plan that is prepared.

The LEPC also receives other information, such as MSDSs and inventory forms, that is submitted by local facilities. It must make this information and the emergency response plan available to the public upon request and establish and publicize procedures for handling such requests. The LEPC has the authority to request additional information from facilities for their own planning purposes or on the behalf of community members. Civil actions can be taken against facilities that fail to provide information required under EPCRA.

The LEPC is also expected to have more "informal" responsibilities within the community. The intention of the regulations are to have the LEPC serve as a "focal point" in the community for information and discussions about hazardous substances and emergency planning and response. Community members are encouraged to go to the LEPC for answers to questions about chemical hazards and potential risks from releases in potentially affected areas. LEPCs are also charged with educating the public about chemical risks and providing them with the means to understand information about the hazardous chemicals and substances used by facilities within the community.

Facilities are required to notify the LEPCs and the SERCs if there is a release of a hazardous substance into the environment in excess of the reportable quantity. The notification activates the emergency response plan and must include, among other requirements, information pertaining to anticipated acute or chronic health risks associated with the chemical release and proper precautions, such as evacuation or medical attention, for potentially exposed individuals in the community.

Public Availability of Community Right-to-Know Information

As discussed, there are two separate community right-to-know reporting requirements involving the use of hazardous chemicals or substances. The first reporting requirement involves either the provision of MSDSs or list of chemicals to the LEPC, SERC, and local fire department. The second reporting requirement involves the annual submittal of a chemical inventory form to the LEPC, SERC, and local fire department. Hazardous chemicals that are present at a facility at any time during a given year above certain threshold quantities must be reported on the inventory form. The regulations provide for a two-tiered approach by allowing the reporting of information about chemicals in applicable hazard categories (Tier I report form) or information about specific chemicals (Tier II report form).

The purpose of these reporting requirements is to provide information to the public about hazardous chemical amounts, types, and locations within the community. The data collected are essential for the planning activities of the LEPC and response personnel. The LEPC must make this information available to the public, and any community member can obtain an MSDS for a specific facility by submitting a written request to the LEPC. Additionally, if the LEPC does not have the requested MSDS in its possession, it must obtain it from the subject facility. The LEPC generally retains MSDSs and other pertinent information at a designated location that is open to the public during normal working hours.

Although facilities have the option of providing either Tier I or Tier II forms, it is the Tier II form that provides more useful information for emergency planning and response, because it applies to specific chemicals rather than hazard categories. The EPA urges facilities to submit Tier II forms that, in any case, must be provided upon request by the SERC, LEPC, or local fire department. Individuals can also request a facility's Tier II information by submitting a written request to the LEPC. If the LEPC does not have a Tier II form in its possession, it must request the form from the owner or operator of the facility. The request must be made by a state or local official acting in his or her own capacity; otherwise, the request is limited to hazardous chemicals stored at the facility in excess of 10,000 pounds. If neither of these conditions is met, the request can still be made by the LEPC if a "statement of need" is included. This explains or justifies the reason for the request.

While MSDS and Tier II forms are available to the public, owners or operators can request that the locations of specific chemicals within the facility be kept confidential. This means that the SERCs, LEPCs, and local fire departments would have access to the location information for emergency planning purposes but could not disclose it to the public.

Owners and operators can also claim that the identity of a chemical is a trade secret and modify their reporting requirements. A valid trade secret claim can, under certain circumstances, protect the name of the hazardous chemical used. However, even if the chemical identity information can be legally withheld from the public, it must be disclosed to health professionals for diagnostic and treatment purposes in emergency situations. In non-emergency cases, a confidentiality agreement must be signed by the health professionals, along with a written statement of need. Information claimed as a trade secret and substantiation for that claim must be submitted to the EPA. Any individual can challenge the trade secret claim by petitioning the EPA. The agency must then review the claim and rule on its validity.

Public Availability of Toxic Chemical Release Information

Certain owners or operators are required to submit yearly reports on the amounts of toxic chemicals that their facilities release into the environment, either routinely or as a result of accidents. The purpose of this reporting requirement is to inform government officials and the public about the release of any of more than 600 listed toxic chemicals to the air, water, land, or injected underground. Facilities subject to this reporting requirement must complete an annual Toxic Chemical Release Inventory Form (Form R), as discussed previously.

The facility uses the Form R report to provide information for each listed chemical and indicate whether it was manufactured, imported, processed, or otherwise used in amounts exceeding the threshold requirements. The regulations require that the owner or operator send the reports to both the federal EPA and to the regulatory agency of the state in which the facility is located. The EPA Office of Pollution Prevention and Toxics is responsible for receiving and processing the forms. The EPA releases a yearly printed report, available to the public, that summarizes the information that was submitted for the annual Toxics Release Inventory.

The EPA is also required to make the data in the reports available to the public through a computer database. The database is intended to provide the public with easy access to information about the presence and release of chemicals in their communities. Any member of the public can access the database and use the information.

The computerized database of TRI data is accessible through various government on-line systems. The National Technical Information Service (NTIS) and the Government Printing Office (GPO) can provide TRI reports via media including magnetic tape, disks, and CD-ROM. The data is also available on microfiche at selected federal depositories and public libraries, and can be purchased from the NTIS and GPO. Additionally, the Right-to-Know Network (RTK NET) contains a collection of databases, including the TRI, that is available to the public via modem and provides information on the reports and a number of related topics.

There are some limitations to the TRI data reported. Some of the data in the inventory can be based on estimates, rather than on actual release measurements. Also, not all manufacturing operations are covered, and the facility must employ more than 10 persons in order to be subject to the reporting requirements. Another limitation is that the annual emissions reports may not indicate whether a chemical was released in large amounts over a short period of time or in small amounts over a longer period of time. Information on the rate at which a chemical is released can provide important risk and health effect information.

The TRI data are used by the EPA, state and local agencies, LEPCs, and other community organizations to gain a broader understanding about the level of chemical usage and release within the community. Environmental agencies use this information to set priorities for further investigation and to determine if possible regulatory or other action needs to be taken to protect public health and the environment. LEPCs and community groups use the information for emergency planning purposes, to learn more about the level of chemical usage in their communities, and to encourage companies to engage in pollution prevention efforts.

CLEAN AIR ACT ACCIDENTAL RELEASE PREVENTION AND RISK MANAGEMENT PROGRAMS

Amendments promulgated in 1990 significantly revised the Clean Air Act, which originally became law in 1970. In particular, the 1990 amendments added a new subsection (subsection 112(r)) on chemical accident prevention. The purpose of this subsection is to prevent the accidental release of regulated substances into the air and to minimize the consequences of such a release, if it does occur, by focusing on preventive measures. The EPA promulgated a list of substances and threshold quantities subject to the accidental release prevention regulations in 40 CFR Part 68. The final rule for implementation of the required risk management program and plan provisions was published in the Federal Register June 21, 1996. Risk management plans (RMPs) for covered sources must be submitted to a central point specified by the EPA prior to June 20, 1999. A general overview of the provisions of the accidental release regulations and risk management program is provided, followed by a discussion of the information that will be made available to the public.

Accidental Release Prevention Provisions

Owners or operators of stationary source facilities that produce, handle, or store substances listed by the EPA in subsection 112(r) above stated threshold quantities are covered by these regulations. Such facilities have what the EPA terms as a general duty to identify hazards that could result from releases, to design and maintain a safe facility, and to minimize the consequences of any releases that may occur. The EPA promulgated a list of more than 100 substances that are subject to these regulations, composed of:

- 77 toxic substances
- 63 flammable substances
- Explosive substances with a mass explosion hazard as determined by the U.S. Department of Transportation (EPA has proposed to delist explosives; thus, explosive substances are not covered under the risk management program and plan regulations)

Threshold quantities established for toxic substances range from 500 to 20,000 pounds. The threshold quantity established for all flammable substances is 10,000 pounds. The threshold quantity established for regulated explosive substances is 5,000 pounds.

The final rule divides the processes subject to these requirements into three tiers, labeled Programs 1, 2, and 3.

The program eligibility criteria are displayed in Table 7.1.

Program 1 is available to any process that has not had an accidental release with off-site consequences in the five years prior to the submission date of the risk management plan and has no public receptors within the distance to a specified toxic or flammable endpoint associated with a "worst-case" release scenario (discussed in the following sections). Program 3

■ TABLE 7.1 Program Eligibility Criteria

Program 1	Program 2	Program 3
No off-site accident history	The process is not eligible for Program 1 or 3	Process is subject to OSHA PSM
No public receptors in worst-case circle		Process is in SIC code 2611, 2812, 2819, 2821, 2865, 2869, 2873, 2879, or 2911
Emergency response coordinated with local responders		

applies to processes in Standard Industrial Classification (SIC) codes 2611 (pulp mills), 2812 (chlor-alkali), 2819 (industrial inorganics), 2821 (plastics and resins), 2865 (cyclic crudes), 2869 (industrial organics), 2873 (nitrogen fertilizers), 2879 (agricultural chemicals), and 2911 (petroleum refineries). Program 3 also applies to all processes subject to the OSHA Process Safety Management (PSM) standard—29 CFR 1910.119—unless the process is eligible for Program 1. Owners or operators need to determine individual SIC codes for each covered process to determine whether Program 3 applies. All other covered processes must satisfy Program 2 requirements.

Risk Management Program Requirements

The different elements required in Programs 1, 2, and 3 are illustrated in Table 7.2.

The owner or operator of a facility whose operations include a covered process must:

1. Prepare and submit a single risk management plan (RMP), including registration that covers all affected processes and chemicals.
2. For Program 1—conduct a worst-case release scenario analysis, review accident history, ensure that emergency response procedures are coordinated with community response organizations to determine eligibility for Program 1 and, if eligible, document the worst-case scenario and complete a Program 1 certification for the RMP.
3. For Program 2—conduct a hazard assessment, document a management system, implement a more extensive (but still streamlined) prevention program, and implement an emergency response program for Program 2 processes.
4. For Program 3—conduct a hazard assessment, document a management system, implement a prevention program that is fundamentally identical to the OSHA PSM standard, and implement an emergency response program for Program 3 processes.

Following is a brief discussion of the hazard assessment, prevention program, and emergency response program elements.

Hazard Assessment   The hazard assessment is intended to evaluate the potential effects of an accidental release of a regulated substance present

■ TABLE 7.2 Program Requirement Comparison

Program 1	Program 2	Program 3
Hazard Assessment	**Hazard Assessment**	**Hazard Assessment**
Worst-case analysis	Worst-case analysis	Worst-case analysis
5-year accident history	Alternative release	Alternative release
	5-year accident history	5-year accident history
Management Program	**Management Program**	**Management Program**
	Document system	Document system
Prevention Program	**Prevention Program**	**Prevention Program**
Certify no additional	Safety information	Process safety
step needed	Hazard review	information
	Operating procedure	Process hazard
	Training	information
	Maintenance	Operating procedures
	Incident investigation	Training
	Compliance audit	Mechanical integrity
		Incident investigation
		Compliance audit
		Management of change
		Pre-startup review
		Contractors
		Employee participation
		Hot work permits
Emergency Response Program	**Emergency Response Program**	**Emergency Response Program**
Coordinate with local	Develop plan	Develop plan
responders	and program	and program
Risk Management Plan Contents	**Risk Management Plan Contents**	**Risk Management Plan Contents**
Executive summary	Executive summary	Executive summary
Registration	Registration	Registration
Worst-case data	Worst-case data	Worst-case data
5-year accident history	Alternative release	Alternative release
Certification	data	data
	5-year accident history	5-year accident history
	Prevention program	Prevention program
	data	data
	Emergency response	Emergency response
	data	data
	Certification	Certification

at the facility above the threshold level. It must include an estimate of potential release quantities and downwind effects, including potential population exposures. The assessment also needs to provide a history of releases that have occurred over a five-year period and include the size, concentration, and duration of each release.

The facility's hazard assessment must include a range of possible release scenarios, including a *worst-case release*, and provide an analysis of potential off-site consequences. The worst-case release is defined as the release of the largest quantity of a regulated substance resulting from a vessel or process line failure, including administrative controls and passive mitigation that limit the total quantity involved or release rate. For most gases, the worst-case release scenario assumes that the quantity is released in 10 minutes. For liquids, the scenario assumes an instantaneous spill. For flammables, the worst case assumes an instantaneous release and a vapor cloud explosion. One worst-case scenario is defined to represent all toxics, and one worst-case scenario is defined to represent all flammables held above the threshold limits at the facility. Additional worst-case scenarios must be analyzed and reported if such a release from another process at the facility could potentially affect other or additional receptors.

In addition to the worst-case release scenario, *alternative release scenarios* must be analyzed. For alternative scenarios, the sources may consider the effects of both passive and active mitigation systems. The non-worst-case releases are presumed "more likely to occur" and "more realistic" than the worst case. The EPA expects owners and operators to select those scenarios that are most useful to communicating with the public and emergency responders about the kinds of release events that are "more likely to occur" at the facility. One alternative release scenario is required for each toxic substance and one scenario is required to represent all flammable substances present above the threshold quantities in processes covered at the facility.

Once the worst-case and alternative release scenarios have been identified, their potential off-site consequences must be analyzed. The off-site consequence analysis uses models or other approaches in order to determine the possible rate, quantity, and duration of a release. The possible distance that a release could travel in any direction before being dispersed to a level where it no longer poses a threat must also be evaluated. The hazard analysis needs to identify all populations that could be affected by the release, including sensitive populations, such as schools and hospitals, and detail potential environmental receptors.

Prevention Program   The risk management plan for the facility must document a prevention program for Program 2 and 3 processes. Program 2 prevention programs must include safety information, hazard review, operating procedures, training, maintenance, compliance audits, and incident investigation. Most Program 2 processes are expected to be relatively simple and located at smaller businesses so that accidents can successfully be prevented utilizing a program that is not as detailed as the OSHA PSM standard requires.

The Program 3 prevention program includes the requirements of the OSHA PSM standard with only a few minor adjustments for program purposes. This was done intentionally, to make it possible for one accident prevention program to protect workers, the general public, and the environment, and also satisfy both OSHA and the EPA. The EPA anticipates that facilities whose processes are already in compliance with the OSHA PSM standard will not need to take any additional steps or create any new documentation to comply with the required Program 3 prevention program.

Emergency Response Program   An emergency response program and plan also need to be developed. The plan required for Program 2 and 3 processes must define the specific actions to be taken in response to a release in order to protect human health and the environment. The emergency response plan incorporates provisions necessary for informing the public, local agencies, and health care professionals in the event of a release. The Program 1 requirements involve only coordination of emergency response activities with local responders.

Public Availability of Risk Management Plans

The risk management plan (RMP) is intended to be a multipurpose document. The RMP must first be utilized to demonstrate compliance with the regulations and must also include the required hazard assessment, prevention program, and emergency response program. The RMP is also intended to provide information to the public in a form that is understandable, and encourages the public to use the information to improve the dialogue with facilities on issues related to release prevention and emergency preparedness.

While all parts of the RMP will be made available to the public, the executive summary is expected to provide the format necessary to understand the various plan components and text descriptions. The executive summary also provides the preparer with the opportunity to explain the facility's programs in a format that will be easy for community members to read and understand. RMPs must be submitted to a central point specified by the EPA prior to June 21, 1999.

At the time this book was written, the EPA was seeking input from stakeholders regarding methods for submitting RMPs so that this information can be made available to all interested parties. The EPA favors the concept of having RMPs available electronically in a database format that anyone can download and search. States, communities, trade associations, or public interest groups may want to use the data (or a data subset) to create databases that allow them to compare information on facilities in the same industry or same area. For example, a local organization could download data from all reporting facilities that are similar to ones in its community to determine whether quantities of chemicals stored or process controls used are typical. The information is intended to provide the public with data that can be used to create or enhance communication with facilities in the community. It is also intended to assist facilities and trade associations in understanding practices in their industries and to identify practices that could be used to reduce risks.

The EPA has not mandated any regulations specifically related to public involvement requirements. Although the agency encourages facilities, the public, and other organizations to work together on accident prevention issues, the wide variety and large number of sources subject to the rule make any single mandatory approach to public participation inappropriate. The RMP information is expected to be used as the basis for a dialogue between the community and facilities on accidental release prevention, risk reduction, and emergency response preparedness. The LEPCs are expected to be used as the conduit through which this dialogue is conducted. Such information about hazards in the community can—and, the EPA believes, should—lead public officials, community members, and facility owner or operators to work together to prevent accidents.

CHAPTER

8

SPECIAL CASES IN COMMUNITY RELATIONS

We have discussed some of the more important concepts in establishing a community relations dialogue: clear, two-way communication and the establishment of positive relationships. We have also discussed the importance of identifying all stakeholders and potential stakeholders and determining how best to communicate with them. These concepts remain constant, regardless of the issues at hand or the circumstances under which a community relations program is undertaken.

This final chapter addresses certain circumstances under which a community relations program may be implemented. It also examines the requirements for federal facilities undergoing investigation and remediation under their Environmental Restoration Program. Finally, this chapter discusses the environmental justice issue. The community relations approach outlined in this book is designed with an eye toward seeking out all potential stakeholders to ensure that, in fact, all affected groups are encouraged to join the dialogue about environmental issues so that no single group has to shoulder a disproportionate share of environmental or health burdens.

COMMUNITY RELATIONS FOR FACILITIES

Most facilities are fixed in one location. In most cases, they have neighbors, either by choice or by chance; need to use public roadways and resources; and interface with public officials. As we discussed in chapter 1, facility management is increasingly being faced with responsibility for disclosing information about operations and environmental performance and for working with stakeholders when making certain facility related decisions.

The following sections discuss two issues in which facility management typically must interact with stakeholders—when undergoing siting or expansion and when disclosing information regarding emissions, discharges, or chemical storage. Other common situations in which facilities may need to interact with stakeholders—in regard to RCRA permitting or in the event of a spill or accident—have been covered elsewhere in this book.

Facility Siting and Expansion

In many jurisdictions within the United States, the siting of manufacturing or waste management facilities in a community requires public "permission" in the form of zoning variances, notification to landowners within a certain distance of the site, or actual siting ordinances. As such, residents have a say, often through a local zoning board, siting committee, or other government unit, if not in person, as to whether a particular land use is acceptable for a given area. Certain facilities that manage wastes must also go through a rigorous RCRA permitting process in order to obtain an operating permit. Although manufacturing and waste management facilities are successfully sited all the time, there are also many instances in which vigorous community opposition either dooms the siting effort or ends in "victory" at the expense of good relations with the community.

Concerns about building new facilities—or expanding existing ones—may include issues relating to the perceived or likely impact on human health and the environment from:

- Increased traffic or changes in traffic routes
- Natural resource use (e.g., withdrawal of substantial quantities of groundwater from an aquifer);
- Changes in the "character" of a neighborhood; or
- Increased emissions or discharges.

As we discussed in chapter 2, the relevance of any of these effects or perceived effects will vary from community to community. For example, residents of areas where traffic is an issue are likely to take a much different view of a facility that will receive a steady stream of truck traffic all day than residents in areas where truck traffic can be routed away from residential neighborhoods or busy sections of town.

Additionally, concerns about natural resource use will also be closely tied to the community's experience with such resources. Availability and use of groundwater and surface water become pivotal issues in the siting or expansion of manufacturing facilities—and in the siting or expansion of waste management facilities, such as landfills—in parts of the country where these resources are limited. Concerns about natural resources also come into play when persons who are involved in hunting or fishing, either for pleasure, commercially, or on a subsistence basis, perceive that the siting or expansion of a facility could have an adverse effect on wildlife or fisheries.

Concerns about changes in the "character" of a neighborhood can be particularly hard to catch before efforts to site a facility are under way. This is especially true in the case of first-time siting efforts. In such cases, residents may never have articulated their feelings about what they would or would not tolerate in their neighborhood simply because the issue has not come up. This is one of the many reasons why a comprehensive community assessment should be conducted before efforts to site a facility are under way.

Much of the "character" issue has as much to do with what residents of a neighborhood want the neighborhood to be as with what it currently looks like. In chapter 2, we discussed the often considerable differences of opinion that can arise among "old" residents and "new" residents. One of the primary differences can be residents' expectations and desires for the area in which they live. Thus, organizations that are considering siting or expanding a facility in or near a residential area need to carefully assess their neighbors and ascertain what they believe the neighborhood is (or should be) aspiring to.

Concerns about emissions or discharges can be triggered in areas with very little industry or in areas with heavy concentrations of industry. For example, some "greenfield" sites can face considerable opposition from neighbors who are concerned about how such areas should be developed and the impact that emissions or discharges could have on the site and nearby properties. Conversely, some of the residents who can be the most critical of the siting of a new facility or the expansion of an existing one are those who live in areas that are already burdened by industrial pollution. Facilities that are relatively "clean" may be welcome in such communities, particularly if they bring jobs and their developers or owners are successful in explaining that they will not add to the environmental burden of the area.

Facilities that require RCRA permits—particularly treatment, storage, and disposal facilities—often face an uphill battle in communities that already see themselves as "dumping grounds." Interestingly, residents will often say, during interviews, that they are aware that a particular facility will not add appreciably to area pollution, or, in the case of treatment or storage facilities, that they understand that no waste will be left on-site. However, residents will also say that they still oppose the siting or expansion of the facility because it's something they *can* do. Other area facilities may be "grandfathered" into a neighborhood, and residents have less opportunity to score decisive victories over them. As a resident of one community said, regarding vigorous opposition by a neighborhood group to block a nearby facility's efforts to expand its operation, "I know they aren't the only source of pollution, or even the worst, but they're a good target because we can do something about them."

Facilities, Jobs, and Taxes

Some developers, owners, or operators of facilities fail to see the importance of community relations because they believe that the economic benefit the facility offers to a community will override most, if not all, concerns that stakeholders might have about environmental issues. While the promise of jobs and an increased tax base can be powerful motivators—particularly in communities that are economically disadvantaged—it is a mistake to believe that these benefits will automatically equal community, and especially residential neighbors', acceptance. Although elected officials and business leaders are often concerned about job growth since they tend to look at an area's total economic outlook, it does not necessarily follow that residents will automatically put economic growth above other interests if development or expansion of a facility is perceived to pose a threat, a nuisance, or an unwanted change to the character of the neighborhood. A good community assessment should be able to determine if such perceptions exist.

Facilities seeking to locate in areas with well-diversified economies, where residents work in a wide variety of occupations and industries, are unlikely to find that residents who are concerned about the perceived environmental consequences of a facility will be swayed by the promise of jobs or an increased tax base. If these residents believe that a facility poses a threat to their health or surrounding environment, they are likely to oppose it.

Although relative affluence and job security may minimize the allure of jobs and tax base, developers, owners, and operators of facilities should also be aware that such a lure does not always translate into a welcome for facilities in economically disadvantaged communities, either. Concerns about environmental justice have sensitized many lower-income and/or minority communities to attempts to site facilities that would not be readily accepted in more affluent communities. Leaders in these communities are taking a hard look at whether the benefits of such facilities truly do override the perceived disadvantages. Many community leaders are beginning to expect much more in the way of a partnership from waste facilities or manufacturing facilities in order to grant permission for siting.

In one case, the developer of a waste treatment facility stated that the facility, which it was attempting to site in a low-income community, would create about 100 jobs. Several residents of the community did some home-work on these jobs and concluded that the majority of them would be going to highly trained technical staff with at least four years of college. Few residents in the community had such credentials—and those who did were already employed elsewhere. Many of the community's residents voiced crit-icism of the developer for failing to disclose that the majority of these jobs were unlikely to go to residents.

Disclosure of Emissions and Discharge Data and Chemical Storage Information

As we discussed in chapter 7, the Emergency Planning and Community Right-to-Know Act (EPCRA) was responsible for making a substantial amount of information about facility emissions and chemical storage avail-able to the public. The availability of this information in the late 1980s marked the beginning of many facility managers' awareness of the need to communi-cate with their communities.

The public availability of emissions data is criticized by some industry representatives because it provides no context about how facilities are actu-ally managing toxics, or the extent to which emissions are (or are not) having a measurable impact on the environment. In fact, it is up to individual facili-ties to provide this information to persons or groups who may be interested or concerned.

Activist groups have used information on emissions and chemical stor-age to push for pollution prevention or, in some cases, toxics use reduction. Facilities that have significant emissions or that store large quantities of chemicals are good targets for these efforts, and activists can often garner significant community support for their activities based on large emission numbers. In some cases, community concerns have encouraged companies to engage in their own pollution prevention or waste minimization initiatives— although, by now, most companies are doing so anyway, for economic as well as health and safety reasons.

However, there are many processes for which less hazardous substitutes are not available, or that create a certain level of emissions despite the use of best available control technologies. The authors have found that, even in cases in which facilities cannot reduce their overall emissions or chemical storage, or can reduce them by only a small amount, community residents *will* often accept current levels of emissions or types of chemicals used if they are provided with explanations regarding the operational constraints on reductions and information on how the current operations are controlled to protect human health and the environment. Concerns about chemical storage can often be dispelled through facility tours, or by explaining the safeguards used to ensure that chemicals are managed safely.

The authors believe it is likely, however, that some facilities may face significant pressure to reduce the amounts of chemicals that are subject to risk management plan reporting (e.g., "worst-case scenario" under Section 112(r) of the Clean Air Act). It is also likely that the disclosure of the risk management plans by the estimated 70,000 facilities that will be required to provide this information will result in community residents' inquiries of nonreporting facilities, as well, as residents are sensitized to the concept of off-site consequences from accidental chemical releases.

Proactive disclosure of emissions data—and of the upcoming risk management plan information—is highly recommended, since it provides facility managers with a greater degree of control over the way information is presented to the public than responding to activists' inquiries or accusations. Such proactive disclosure works best, however, if relationships have already been established and are being maintained. This is especially true in the case of the risk management plans under Section 112(r). One community resident, upon hearing about the upcoming risk management plan disclosures, commented to the authors that introducing a facility to the community for the first time through its risk management plan was like introducing someone to electricity through the electric chair—not only would it be a shocking experience, it would create a very negative impression.

Investigation and Cleanup of Contaminated Sites

The presence of contaminated sites can create serious concerns among both residential and commercial neighbors. If groundwater that is used as a drinking water supply has been affected, or if runoff from the site is affecting surface water that is used for drinking water, recreation, or fishing, then a significant number of people may be concerned. Similarly, contaminated sites that are seen as hampering economic development or that are viewed as stigmatizing a community or neighborhood can also create concerns beyond those involving health. In such cases, the site is essentially viewed as a wound to the community.

Community residents' concerns regarding contaminated sites typically center around health, especially increased cancer risk or reproductive problems, and the effect the site may have on property values. These concerns are frequently exacerbated by the extended time frames of investigation and remediation activities at many sites. Investigations conducted under CERCLA or state-run programs, in particular, must proceed through standard steps in the investigation and cleanup process. Each step involves the preparation of documentation, such as work plans or reports of investigative findings, and review of these documents, not to mention the time required for collecting samples, having them analyzed at a laboratory, and evaluating the results. Complex sites may require several iterations of sampling and analysis (a "phased approach") to characterize the contamination and the underlying geology well enough to design a remediation strategy. Thus, investigation and remediation of a site can literally take years to complete.

The authors have found that not only can providing stakeholders with information about how an investigation is conducted minimize some of the anxiety and anger over the length of time involved, but this information can also enhance stakeholder understanding of the site's condition—which can assist in subsequent discussions about remediation alternatives as they are developed. With this in mind, persons responsible for community relations in conjunction with large sites that are the subject of significant community concerns should prepare and distribute fact sheets and newsletters to keep stakeholders up to date on activities at the site and convey a sense that investigative or cleanup activities are moving forward. The time frame issue can also apply to minor, voluntary action sites where investigations are actually proceeding very rapidly according to technical standards—but slowly in the eyes of the neighbors.

If contaminated sites are visible to residents and conditions are such that workers will need to wear personal protective equipment in addition to the standard gloves, boots, and coveralls required to perform investigative and cleanup activities, then neighbors, officials, and other relevant stakeholders should be briefed to reassure them that people outside the site are safe. Situations in which samples are collected from neighboring residential or commercial properties definitely require that residents, owners, or managers be briefed as to the intent of the sampling effort. Nearby property owners or residents typically express mixed feelings about having samples collected on their property. Fears about the presence of contamination can be alleviated through sampling and analysis; conversely, property owners or residents are also frequently worried about the consequences of confirming contamination on their property. Community relations specialists need to work closely with neighbors during such investigations and make sure that their wants and needs are reflected in the decision-making process.

Cleanups involving residential property can also result in serious concerns. Although the overriding concern would appear to be getting rid of the contamination—and, certainly, residents typically want that—as in the case of investigations on residential property, residents often suffer from considerable anxiety and mixed feelings about remedial activities taking place on their property. Neighbors view such activities as very personal intrusions; thus, the need for empathy and understanding—as well as the planning of remediation activities so that they can be completed as quickly as possible with a minimum of disruption for residents—is a must.

Community Relations and Risk-Based Cleanups   Community relations in conjunction with cleanups to meet risk-based standards will likely require an increased amount of communication with stakeholders. Risk-based cleanups often result in some contamination remaining in the ground, as opposed to restoring the site to "pristine." The cleanup standard can depend on the planned future use of the property—a site that is going to be redeveloped for heavy industrial use, for instance, would be required to meet a lower standard than a site intended for some other use. However, given the fact that some stakeholders may feel that no level of contamination is acceptable, significant effort may have to be expended to convince them that the use of risk-based standards are sufficient to protect human health and the environment.

Some of the sites that are likely to be cleaned up according to risk-based standards are so-called brownfield sites—industrial or commercial sites with some contamination from past use that are intended to be redeveloped for reuse as industrial or commercial property. Since many of these sites are located in low-income or minority areas, managers responsible for community relations should be sensitive to environmental justice issues tied to the use of risk-based standards. Care should be taken to ensure that area residents and other stakeholders are fully engaged in a dialogue about the site with those responsible for cleanup and redevelopment so that their questions and concerns are adequately addressed.

Nonenvironmental Issues Masquerading as Environmental Concerns

The authors, and others who work in the area of environmental community relations, periodically encounter situations involving concern about the environment that are, in fact, either a façade for pursuing another agenda or simply a facet of another conflict or problem. We've written at length about how other, nonenvironmental, issues can influence stakeholders' concerns or perceptions. In the cases we are discussing in this section, however, the environmental issue is really incidental to the other issues in question—even though it may appear at first glance to be the key issue.

In the authors' experience, nonenvironmental issues masquerading as environmental issues surface most often in conjunction with labor disputes and land use decisions. In both of these cases, persons who are seeking to attain another goal (e.g., more power for a union or discouraging development of property) play the "environmental card" as a ploy to achieve other goals. In doing so, however, they frequently sell others in the community on the validity of the environmental concern they have raised. These other people typically have no idea that the concern was raised, in large part, to pursue another goal; thus, their concerns need to be answered in the same way as any other environmental concern. It is not always possible, for political or business reasons, to bring "ulterior motives" out on the table. Regardless of the root causes of the environmental issues that are raised, though, the issues need to be dealt with at face value.

Persons who oppose the development of property—or the type of development proposed—often find that they can garner support by raising environmental objections. In one such case, neighbors of a vacant parcel of land opposed the development of an office park, citing the potential for damage that runoff from the parking lot would cause to a nearby lake. Thanks to grassroots support, the neighbors won their case at the zoning board level. The runoff problem apparently disappeared entirely when, several years later, one of the neighbors (who had since bought the property) built a shopping center on it.

Playing the "environmental card" over land use is also frequently encountered in regard to transportation projects. Such situations can be very complicated to address, however, because the building of roadways does affect the environment—from the construction activities themselves, increased development at interchanges, and increased air emissions from traffic. Thus, separating real environmental issues from those raised that have more to do with perceived impact on quality of life can be difficult. As we have attempted to convey in this book, it is best, in such situations, to address the environmental concerns that have been raised while acknowledging the other nonenvironmental concerns, rather than dismissing the environmental aspects of the issue out of hand because they may, in fact, be secondary.

There are also situations in which the "environmental card" is played on a more oblique and unconscious level. In one such situation encountered by the authors, employees raised concerns to management about suspected contamination at their facility. Although it was clear that the employees believed that there could be a problem, their escalation of the issue and the very combative atmosphere that developed over the way the investigation was conducted were firmly rooted in a conflict between the union and the facility's management.

FEDERAL FACILITIES PROGRAMS

Current estimates indicate that the U. S. government is responsible for more than 61,000 contaminated sites or facilities throughout the United States. Affected sites range from abandoned mines to major weapons production centers, and the government estimates that between $230 billion and $390 billion will be spent over the next 75 years for environmental remediation and restoration. The Federal Facilities Environmental Restoration Program was established in order to address the identification, evaluation, and remediation of these sites and facilities.

According to the Federal Facilities Environmental Restoration Dialogue Committee, a group chartered by the EPA and composed of representatives drawn from federal agencies, state, tribal, and local governments, and other environmental, community, environmental justice, and labor organizations, most of the environmental contamination associated with federal facilities is the result of Department of Defense (DOD) and Department of Energy (DOE) activities. Many of these facilities were involved in serving national security interests, and some continue to function in that capacity. Conversely, the U.S. Department of Agriculture (USDA) and Department of Interior (DOI) are believed to be responsible for only a small portion of the contamination found on government lands. These departments are responsible for the stewardship of federal lands where contamination has resulted primarily from mining activities, pipeline breaks or petroleum product releases, and landfilling operations. Additionally, illegal activities such as "midnight dumping" have contributed to the level of environmental contamination found on federal lands that are under the authority of the USDA or DOI.

Because of the wide variety of affected sites, facilities, and locations, the government is particularly concerned with issues related to community involvement and environmental justice. Sites and facilities are found in locations that range from isolated mining areas in the west, to major population centers, to the lands of Native Americans. In many areas, the economic well-being of the surrounding community is closely tied to the outcome of the cleanup effort and subsequent decision on how the remediated land should be used. Because each location and type of site presents a unique set of circumstances and the investment of a considerable amount of time and money, the government has taken a very proactive approach to including stakeholders in the decision-making process.

The following discussion presents an overview of the Federal Facilities Environmental Restoration Program and the community relations activities that the government is implementing as an integral part of the program for the subject sites and facilities.

Overview of Program

The Superfund Amendments and Reauthorization Act (SARA) brings federal facilities under the jurisdiction of the Comprehensive Environmental Response, Compensation, and Liability Act (CERCLA). The National Contingency Plan (40 CFR Part 300) contains the implementing regulations under SARA, and the federal facilities restoration program is charged with conforming to these provisions. Regulatory requirements for environmental community relations programs are discussed in chapter 7.

A recent dialogue committee report indicates that past public involvement approaches associated with DOD and DOE contaminated facilities often resulted in a significant degree of mistrust among stakeholders. The national defense missions of these facilities often meant that very little information could be released for public consumption, and the information that was made available was often so excised of meaningful content that it was essentially worthless. Additionally, efforts were seldom made to include stakeholders, particularly those in minority and economically disadvantaged communities, in discussions about the planned remediation efforts or future land use. Also, in many cases, local governments were not informed or adequately involved in the decision-making processes taking place in their own communities.

Because of the magnitude of the contamination and number of sites and facilities affected, the government has needed to set priorities for federal cleanup activities. Remediation of contamination at DOE and DOD sites is publicly funded. However, the USDA and DOI must look to responsible parties to conduct cleanup activities at the sites or facilities where they contributed contamination or, if the responsible parties cannot be located, seek to recover public funds that have been set aside (such as through CERCLA) for remediation efforts. In cases in which DOE or DOD activities resulted in contamination of USDA or DOI managed lands, funds must be sought from the DOE and DOD in order to pay for the cleanup costs.

Several guidances and manuals, some authored by the dialogue committee, have been published to assist government agencies and contractors in dealing with the unique problems and concerns associated with the restoration of federal sites and facilities. These guidances provide the fundamental principles and a framework for the government to use in addressing the complex issues arising from environmental contamination at these sites and facilities. The framework includes maintaining community involvement programs, establishing stakeholder advisory boards, funding and priority setting, and enhancing the capacity of all stakeholders to participate in the decision-making process.

Principles for Environmental Cleanup of Federal Facilities

The committee released a guidance providing 14 principles that comprise the basis for making federal facility cleanup decisions and apply to all persons and institutions involved in the process. The principles are presented in the box on pp. 317 and 318.

Community Involvement

Community involvement is a vital part of the programs undertaken to remediate environmental contamination and restore federal sites and facili-

Principles for Environmental Cleanup at Federal Facilities

1. Nature of the obligation—the federal government has a legal, ethical, and moral obligation to clean up the contamination in a way that, at a minimum, protects human health and the environment and minimizes burdens on future generations.

2. Sustained commitment to environmental cleanup—to complete cleanups at a reasonable and defensible pace that is protective of human health and the environment and allows closing federal facilities to return to economic use as soon as possible.

3. Environmental justice—special efforts must be made to reduce negative impacts of contamination related to federal facility activities on affected communities that have historically lacked economic and political power, adequate health services, and other resources.

4. Consistency of treatment between federal facilities and private sites—federal facilities should be treated in a like manner to private sites, particularly in terms of the application of cleanup standards.

5. Cleanup contracting—contracts should be managed as efficiently as possible and include opportunities for local communities.

6. Fiscal management—funding mechanisms should be flexible, in timing of expenditures, and efficient.

7. Interdependent decision-making roles and responsibilities—numerous institutions and people play very distinct roles in the decision-making process for cleanups. Process must ensure that all roles are preserved and balanced.

8. The role of negotiated cleanup agreements—they are critical in setting priorities and providing a means to balance roles and responsibilities in decision making.

9. Consideration of human health risk and other factors in federal facility environmental cleanup decision making—risk assessments are primary tool used to evaluate risks to human health but have recognized limitations. Other factors warrant consideration, including: cultural, social, and economic factors and environmental justice; short- and long-term ecological effects and environmental impacts; land use; acceptability to regulators, tribes, and public stakeholders; statutory requirements and legal agreements; life-cycle costs; pragmatic considerations; and overall cost and effectiveness.

10. The importance of pollution prevention and control activities—essential to prevent future environmental problems.

11. The role of future land use determinations in making cleanup decisions—should be considered, provided that, at the time of land transfer, there are adequate safeguards to protect landholders, lessors or receivers, and surrounding communities. Communities, local governments, and affected Indian Tribes should be given significant roles in determining future use.

12. The role of studies in the cleanup process—efforts to streamline process focus on reducing paperwork and moving away from adversarial relations, not arbitrary capping of funds for studies, which are considered important.

13. Need for systematic approach for decision making and priority setting—priority setting decisions should be made in a manner that recognizes their interconnectedness to other environmental problems.

14. Stakeholder involvement—steps must be taken to ensure stakeholder involvement with the goal of achieving cleanup decisions and priorities reflecting a broad spectrum of stakeholder input from affected communities including indigenous peoples and economically and socially disadvantaged groups.

From *The Final Report of the Federal Facilities Environmental Restoration Dialogue Committee*, April 1996.

ties for future use. As mentioned, in the past, stakeholder involvement was often an inconsistent and ineffective part of the restoration programs, particularly at DOD and DOE facilities. The community involvement programs that were instituted often did not reach out to all affected stakeholders.

A number of steps have been taken to enhance public involvement at federal site and facility restoration projects. Field staff are encouraged to conduct community assessments of stakeholder needs and concerns before initiating community involvement programs. These assessments are intended to ensure that the program developed is appropriate to the specific community and that resources are not wasted with respect to the public involvement efforts and community relations activities chosen. Agency policies direct field staff to actively reach out to, and solicit input from, all of the stakeholders in the community.

Additionally, diverse methods of communicating with the community are encouraged, including:

- Utilizing local media outlets such as television, local cable channels, newspapers, and Internet services, to increase public awareness and involvement
- Ensuring that materials for public participation are culturally sensitive and relevant to the socioeconomic area—including providing information in appropriate languages and level of detail.
- Using local government and other institutional community involvement mechanisms for information exchanges regarding the cleanup efforts.
- Designating information repositories at appropriate and convenient locations.
- Applying Freedom of Information Act (FOIA) exemptions narrowly, so that as much information as possible is made available to the public.

Community involvement is also an integral component to the successful determination of future land use at restored federal sites and facilities. The decision made concerning the most appropriate use for the federal land after cleanup is often critical to the economic outlook of the affected community. Local government officials and community members need to be intimately involved in this process.

The DOD, DOE, USDA, and DOI have each developed policies and procedures specifically designed to enhance and encourage stakeholder involvement in the restoration program.

DOD Public Involvement Policy The DOD restoration program policy emphasizes the need for providing early access to relevant information so that the public is adequately informed about the process and has the opportunity to provide input during the planning stages. Specifically, the policy requires installations to make the following documents available to the public:

- Draft final and final technical documents;
- Proposed and final plans; and
- Status reports.

In addition, DOD documents that are considered to be regulatory agency "deliverables" must be made available to the public at the same time they are provided to the agency.

DOD installations are required—through regulations, as part of the CERCLA program—to establish an information repository at a central, accessible, and convenient location. The installations also maintain mailing lists that are used to contact community members who have expressed an interest in the program. Fact sheets and other materials can then be easily and routinely distributed to persons on the mailing list.

Many installations have themselves developed a policy of producing brief summaries of documents for distribution to community members in an effort to make the very technical reports produced during the investigation and remediation process more readable and easier to understand. Technical personnel are increasingly available at public meetings and other designated forums to ensure that public concerns and questions are answered. In fact, many installations have established "points of contact"—persons who are charged with providing information to the community about the restoration program and activities. These contact persons work with the community to ensure that there is a swift and adequate response to their questions. The contact persons also have the authority to refer the public to other sources, if necessary, to obtain information.

The DOD is creating regional environmental coordination offices to facilitate environmental functions among the services. A regional environmental coordinator will ensure that issues are brought to the attention of the appropriate personnel in an attempt to solve some of the past information distribution problems and concerns.

DOE Public Involvement Policy The DOE's public participation policy commits the department to information exchanges and an ongoing two-way community dialogue. The DOE's policy statement includes a commitment to conducting all formal and informal public participation activities "in a spirit of openness, with respect for different perspectives and a genuine quest for a diversity of information and ideas." The policy also commits the department to establish, announce, and manage topical databases of reliable, timely information made available to the public via telephone and computer. The Office of Environmental Management at the DOE has established an information center intended to provide public access to all program material. The center maintains an 800 telephone number, electronic bulletin board with e-mail access to all DOE employees, and an extensive library.

Each major DOE site has appointed a public participation coordinator, who serves as the central point of contact for all public participation activities. The coordinator is responsible for ensuring that the public has meaningful opportunities to participate in policy formulation and decision making with respect to restoration program activities.

The Office of Environmental Management has also established the Office of Public Accountability to coordinate public participation efforts on a national level. This office operates a training program for senior and mid-level managers, oversees an information center, manages 11 site-specific advisory boards, and maintains a national dialogue with state and tribal leaders.

DOI and USDA Public Participation Policies The DOI bureaus are responsible for distributing information to the public and establishing central points of contact for their programs. Each bureau has an office, typically at the headquarters location, that is assigned primary responsibility for the program. Additionally, the Office of Environmental Policy and Compliance serves as an information clearinghouse and contact point, both for headquarters and for the eight regional offices.

USDA agencies with relatively small programs—for example, the Agricultural Research Service—designate an individual to act as environmental pollution control coordinator. This person is designated at the headquarters level to be the contact point for program activities. The USDA-Forest Service policy requires each region and research station to designate an individual as a CERCLA coordinator and to serve as the primary contact point with headquarters. The CERCLA coordinator is responsible for the organization of all aspects of the response actions initiated under CERCLA. Responsibilities for specific projects are typically delegated to an on-scene coordinator or remedial project manager. The public affairs specialist is the designated spokesperson for community relations activities during response actions.

Advisory Boards

Advisory boards provide one method of formally establishing the community's decision-making role in the restoration program. The advisory boards are used to complement, rather than duplicate or replace, other site and facility restoration-related public involvement activities. Current estimates are that more than 200 advisory boards have been established specifically to address federal facility cleanups—the DOE has established 11 site-specific advisory boards (SSABs) and the DOD has established more than 200 restoration advisory boards (RABs).

The DOE's Environmental Management program was responsible for establishing 11 boards in cooperation with state environmental agencies and regional EPA offices. The SSABs are designed to provide advice on site-specific issues; however, they also meet several times a year on an ad hoc basis to discuss both site related and national environmental issues and concerns. A national environmental management advisory board has also been established. The goal of this board is to provide program advice on a national level to complement that provided by the SSABs. Additionally, the State and Tribal Government Working Group, established in 1989, provides a voice for affected states and Indian nations in the decision-making process.

These advisory boards have served to dramatically improve the relationship building and decision-making processes at a number of facilities.

For example, one of the more successful public involvement initiatives in the DOE's nationwide network of nuclear materials facilities occurred at the Fernald site near Cincinnati, Ohio. Past relations between Fernald and its neighbors were so strained that 14,000 area residents filed a class action lawsuit against the agency and its prime contractor, seeking damages for exposure to off-site pollution and loss of property values. Even after the lawsuit was settled in 1988, the dialogue between plant operators and neighbors continued to be uneasy, as the mission at Fernald changed from uranium processing to environmental cleanup in 1990. However, the public involvement initiatives that coincided with the Fernald site's mission change have continued to gather support from both agency officials and citizens. A Fernald Citizens Task Force, formed in 1993, was the first formally organized site-specific advisory board in the DOE system and is cited by both citizens and agency officials as an important success story. Among the innovative public involvement approaches at Fernald is a so-called partnering experiment, in which individual members of the citizens' task force are matched with individual site officers and managers to improve communication and expedite problem solving. (From *The Final Report of the Federal Facilities Environmental Restoration Dialogue Committee*, April 1996.)

The DOD has established more than 200 RABs at closing installations and at operating installations where community members have expressed an interest. The RABs are intended to bring together people reflecting diverse interests in the community to facilitate the communication activities and information flow between the affected community, the DOD installation, and regulatory agencies. The RABs are intended to complement other community involvement efforts, not to replace them. The member selection process is designed to be open and is conducted in cooperation with the regulatory agencies and community. RAB community-based membership is composed of persons who live or work in the affected community or could be affected by the restoration program. The installation selects a DOD co-chair and the community members select a community co-chair after the RAB has formed.

The co-chairs serve as equal partners in establishing the RAB's agenda, and are responsible for ensuring that members have the opportunity to raise issues and concerns in a open and constructive manner.

Each installation is charged with providing administrative support to its RAB. This can include the provision of facilities to hold the meetings, preparation of meeting minutes, distribution of mailings and public notices, and facilitation of meetings.

The advisory boards can provide important independent policy and technical advice on the site or facility restoration project and generally improve site related decision making through:

- Providing a setting for regular direct contact between agencies, facilities or sites, and affected stakeholders.
- Providing a forum for understanding the often competing needs and requirements of the government and affected communities.
- Enabling the development of more satisfactory plans through public input.
- Allowing for a more in-depth analysis of technical plans and reports than would be possible in the usual single public meeting.
- Permitting a more detailed consideration of issues than the legal minimum requirements in various laws and regulations.
- Allowing cleanup decisions to consider community-based values and judgments, as well as technical data.

An assessment of existing public involvement programs is usually recommended before initiating the establishment of a board, in order to determine both the interests and needs of the community. The assessment typically incorporates input from local governments, community groups, local citizens, and other stakeholders. Where advisory boards already exist, the intent is to build on the relationships that have already been established. This can be done by increasing the scope of issues addressed, adding members to ensure that all stakeholder groups are included, or enhancing the way the board interacts with the public.

The fundamental purpose of the boards is to ensure the protection of human health and the environment during site and facility cleanup and restoration activities. However, other issues, such as waste management, technology development, employment opportunities, and future land use considerations, may also compose part of the board's mandate.

Funding and Priority Setting

It is essential that the facility and community relationships be established early in the restoration program process to ensure that the funding and priority setting decisions made are acceptable to the community. Additionally, this early involvement is critical for the eventual "buy-in" of the community to the remediation strategy and future land use decisions that are made. Stakeholder involvement is recommended, as much as possible, in decisions regarding the scope, timing, and priority of activities to be performed. In instances in which stakeholders cannot be directly involved in certain funding and budgetary decisions, they should be advised of, and have the opportunity to comment on, the relevant decisions.

Capacity Building

The ability of all stakeholders to participate effectively in the decision-making process is essential to the success of the restoration program. Enhancing the capacity of all stakeholders to participate, in turn, helps establish effective working partnerships among regulatory agencies, the installations, and stakeholders.

The capacity of stakeholder groups to effectively participate in community involvement activities, advisory boards, and budget and priority setting processes is a key concern for program participants. Many state, local, and tribal governments have taken on increased regulatory responsibilities and attempted to expand their capacities as partners in the federal facility restoration program. Federal regulatory agencies have also sought to enhance their abilities to communicate and work with all of the affected stakeholder groups. In particular, there has been a focus on expanding communication channels to low-income and disadvantaged groups.

The dialogue committee's report emphasizes that local governments play an important role in the federal facility cleanup process, which can often stretch or exceed their real capacities and traditional responsibilities. Local governments face added responsibilities because of the complex nature and extensive cleanup efforts required at federal sites and facilities. These can include analyzing and addressing the impacts of federal actions on public safety and health; planning for and responding to facility- or site-related emergencies, such as explosions or fires; ensuring that local sewage systems are capable of handling unique waste streams; and managing transportation routes and ground and surface water resources. They also play a key role in planning for the future land use of the properties.

These additional responsibilities can put a strain on many smaller local governments, particularly those in rural communities. Such communities often do not have full-time paid officials or sufficient technical staff or exper-tise to fully participate in the program. Local governments may need the assistance of state and federal government resources to determine the best mechanisms to support and enhance their capacity to participate in the restoration program in affected communities.

Special efforts also need to be made in certain communities in order to ensure that the public has the capacity to fully participate in the decision-making process. Low-income, minority, and disadvantaged communities may require special assistance in order to develop the technical and analytical expertise necessary to understand the complexities of the restoration pro-cess. Training and technical assistance programs may be developed to reach out to affected communities. Additionally, department agency personnel may need to undergo training in order to comprehend all of the components necessary to successfully reach out to affected stakeholder groups.

For example, the DOE's Office of Environmental Management has insti-tutionalized training in public participation principles for managers, technical staff, and the stakeholders with whom the office interacts. Certain courses are offered to DOE program/project managers and their technical and public participation support staff. These courses are intended to help managers and staff recognize their responsibilities to stakeholders, plan and manage public involvement, and to become personally involved in interacting with stake-holders. Stakeholder training consists of a one-day workshop featuring pre-sentations and class activities designed and presented by DOE stakeholders. The course offers an opportunity for DOE employees and stakeholders to meet, interact, and learn from their respective experiences. (From *The Final Report of the Federal Facilities Environmental Restoration Dialogue Com-mittee*, April 1996.)

ENVIRONMENTAL JUSTICE

Over the past several years, the concept of environmental justice and con-cerns about the effects of environmental pollution on certain segments of the population has gained increased national attention. The environmental jus-tice agenda is driven by concerns that minority and low-income populations bear disproportionately high and adverse human health and environmental effects from pollution. Advocates of environmental justice initiatives believe that environmental laws and regulations have not been applied equally to certain population groups and geographic locations. This disparity in environ-mental protection has resulted in the increased vulnerability of minority and low-income communities to health threats from pollution, which are often exacerbated by other factors, such as inadequate health care resources and poor nutrition.

The following "draft" definition of environmental justice was prepared by the National Environmental Justice Advisory Council (NEJAC):

> The fair treatment and meaningful involvement of all people regardless of race, ethnicity, culture, income, or educational level with respect to the development, implementation, and enforcement of environmental laws, regulations, and policies. Fair treatment means that no population, due to political or economic disempowerment, is forced to shoulder the negative human health and environmental effects of pollution or other environmental hazards.

NEJAC, which was formed by the EPA in 1994 to advise the agency on matters related to environmental justice, is composed of 23 representatives from academia, business and industry, state, tribal, and local governments, environmental organizations, community groups, and non-governmental organizations.

Environmental justice issues have arisen in the context of many federal and state-related facility permitting and siting programs. For instance, recent challenges under Title VI of the Civil Rights Act allege that federal grants allocated to states to support state RCRA permitting programs are being administered in a discriminatory manner. In the past, the EPA has traditionally focused on site-specific and operational factors in determining whether or not to issue a RCRA facility permit. The agency is now exploring additional ways of using risk or health assessments to determine whether the surrounding community would face unacceptable human health or environmental effects if the permit were issued.

The federal government has taken the lead in addressing environmental justice concerns—as it has with many other social and civil rights issues. Executive Order 12898, signed by President Clinton in 1994, establishes environmental justice as a national priority. The order, "Federal Actions to Address Environmental Justice in Minority Populations and Low-Income Populations," directs federal agencies to "identify and address disproportionately high exposures and adverse human health or environmental effects of their programs, policies, and activities on minority and low-income populations." All agency strategies must consider statute enforcement in these areas and also provide greater public participation opportunities, improvement of health-related research, and identification of differential patterns of subsistence use of natural resources. The order is intended to promote nondiscrimination in federal programs and, most importantly, to provide minority and low-income communities with increased access to public information and additional opportunities to participate in decision-making activities related to human health and the environment.

EPA Programs and Strategies

Environmental justice issues present a unique set of challenges for the EPA. Executive Order 12898 essentially mandates the consideration of environmental justice issues in all of the EPA's programs, policies, and decisions. Stakeholder groups from the affected minority or low-income communities need to be identified and involved early in the process in order to ensure that environmental justice issues are considered and addressed before they become problematic. We have discussed regulatory-based environmental community relations programs and public participation requirements in chapter 7. The following discussion presents EPA environmental justice priorities with respect to RCRA and Superfund programs.

RCRA Environmental Justice Issues   Environmental justice issues in regard to the Resource Conservation and Recovery Act (RCRA) program have arisen in almost every aspect of hazardous waste management, from facility siting, to permitting, to corrective actions. The EPA has committed regional and state agencies to conducting environmental justice pilot projects in areas targeted for priority permitting activities. These pilot projects are expected to incorporate a number of aspects, including:

- Increasing public involvement by tailoring outreach activities to affected communities.
- Factoring unique environmental justice considerations into public health surveys or assessments.
- Evaluating demographics, including examining differing population types and income levels at various RCRA facilities.
- Specifying permit conditions that address particular community-based or demographic concerns.

Environmental justice issues also often arise when establishing corrective action priorities for a facility. Although the agency's priority-setting process does not specifically address environmental justice issues, there are instances in which such issues may be used to elevate the corrective action priority of the facility. Additional, and more intensive, public outreach efforts may also be undertaken at facilities where environmental justice issues are raised in relation to the corrective action process.

A recent EPA final rule for expanded public participation, also discussed in chapter 7, was designed, in part, to address environmental justice concerns by increasing public participation requirements in the RCRA permitting process. The regulations promulgated in the final rule significantly increase the level of public participation early in the permitting process—before the permit application is even submitted. These measures are intended to provide minority and low-income communities with a greater opportunity to influence and be directly involved in the decision- making process from the outset.

Superfund Environmental Justice Issues   The Superfund program is focusing on several fronts involving environmental justice concerns and issues. One of the most significant areas involves efforts associated with community involvement and outreach. A major item, which was originally incorporated in the recent effort to amend Superfund, is the establishment of community advisory groups (CAGs) at National Priorities List (Superfund) sites that have environmental justice issues. The CAGs are intended to encourage early community participation in the Superfund process. They are designed to provide input to the agency on all significant decisions pertaining to the site, including proposed remediation alternatives and future land use. CAG membership is expected to represent the composition of the community affected by the site and the diversity of local interests. Draft CAG guidance suggests that at least half of the members should be local residents.

The Superfund program is also studying the use of proactive preliminary site assessments for areas with environmental justice concerns, in order to ensure that problem sites are identified. Risk assessment approaches are also being reassessed with environmental justice issues in mind. The agency hopes to develop tools for site managers to use to factor in multiple exposures and unique risk scenarios with respect to the assessment of risk in minority and low-income communities.

Additionally, the agency is evaluating the remedy selection process and speed of cleanups with respect to environmental justice issues and the potential adverse effects on areas with minority and low-income populations. Population and demographic information is increasingly being used early in the remediation process to ensure the identification of potential areas of concern—hopefully before major environmental justice issues arise.

Environmental Justice Projects and Sustainable Development

Federal, state, and local governments have undertaken and supported the development of a number of model or pilot projects intended to provide information and positive examples for community organizations and private business enterprises to use in addressing environmental justice concerns. Following are two examples of the kinds of projects that are being undertaken with the support of government agencies in an effort to address environmental justice issues. A multidisciplinary approach is used that does not simply focus on the specific environmental project, but also considers other factors, such as employment and economic opportunities, that must be an integral part of an environmental justice strategy.

Brownfields Projects   Government agencies, developers, community organizations, and others have supported a number of major initiatives intended to encourage the cleanup and reuse of abandoned, idled, or misused industrial and commercial facilities located in urban population centers. These are commonly termed "brownfields" projects, and are often located in minority or low-income communities. Opportunities for cleanup and redevelopment of these sites have historically been complicated by a number of factors, including existing environmental contamination and stringent cleanup standards; competition with "greenfield" suburban sites; strict local ordinances and regulations; and other employment, economic, and social issues.

The brownfields initiatives are intended to encourage investors, real estate developers, lenders, community groups, and other affected stakeholders to take a fresh look at how these sites can be redeveloped to bring jobs and economic opportunities back into older urban industrial settings. Brownfields pilot projects throughout the country are being used to find new ways in which the environmental cleanup of sites intended for industrial redevelopment can be streamlined using risk-based standards that are appropriate to the future land use, but not detrimental to the community.

These redevelopment activities are designed to provide a sustainable economic base for the community and thus require the support of all the affected stakeholder groups. Successful brownfields strategies require the intensive involvement of all affected parties, including government agencies, private enterprises, and community members. Communication is typically required to educate and inform all public and private parties about the availability of the brownfields sites and the philosophy behind the cleanup and revitalization efforts. All parties must be aware of the importance of coordinating the activities involved in cleaning up these industrial sites for economic redevelopment.

Community Chemical Emergency Preparedness and Prevention Outreach and Training

The community of Kellog, Idaho, lies within the boundaries of the Bunker Hill Superfund site. This former mining/smelting industry community—with a recent unemployment rate that hovered around 25 percent—found itself in the midst of a multimillion-dollar Superfund cleanup effort. Citizens were concerned that, despite the potential employment opportunities arising due to the cleanup activities, such jobs would not be filled by local community members. Although the EPA encourages its cleanup contractors to hire locally, the problem was that most members of the community had not completed the 40-hour health and safety training required for employment at cleanup sites.

The Region 8 Superfund Site Response Section piloted a program to conduct hazardous waste health and safety training in communities with nearby Superfund sites. A goal in 1993 and 1994 was to target communities with high unemployment rates to encourage the employment of local individuals by cleanup contractors working at the Superfund sites. The program reports that 90 people in Kellog were trained and certified, making them eligible for employment at the Bunker Hill site. (From *The Environmental Justice Strategy: Executive Order 12898*, U.S. EPA, Office of Environmental Justice, EPA/200-R-95-002, April 1995.)

Projects such as this have a twofold goal. One goal is to communicate with local residents in order to increase awareness and knowledge about the environmental projects occurring in their community. This often has a secondary effect of increasing the level of public participation and support for the project. The other goal is to address issues, such as unemployment, that are commonly related to, and often inextricably linked with, environmental justice concerns.

Environmental Justice Initiatives

The principles behind the federal environmental justice strategy can be applied to many kinds of situations. The strategy takes into consideration, by definition, that all groups do not have equal access to environmental protection. The strategy begins from the presumption that certain groups lack access to decision makers and inclusion in the decision-making process. Those groups also often lack the educational tools and technical resources that are necessary to effectively participate in environmental planning. Because of this, the strategy must involve working within communities to communicate directly with affected stakeholders and to engage them in a constructive dialogue.

Communication and Public Involvement Affected communities must have access to information that will enable them to fully participate in activities related to environmental planning and policy implementation. Providing access to information may be as simple as ensuring that notices and fact sheets are available to community members in their language of origin. However, additional communication efforts can include one-on-one meetings, availability sessions, and other public forums specifically designed to discuss environmental justice and other local environmental concerns.

As we have discussed elsewhere in this book, communication efforts can be enhanced by ensuring that all public documents and notices related to human health and the environment are written in clear, understandable terms that are familiar to the majority of community members. Such notices or documents, as well as being written in languages other than English when necessary, need to be composed with the particular sensitivities of the local community in mind. Notices of public meetings should be made available through channels that are accessible and utilized by the affected population, such as local minority-oriented newspapers or non-English radio and television stations.

Environmental justice issues and concerns should also be considered in the planning and execution of public meetings, hearings, or other forums that may be necessitated or required under many different circumstances and environmental programs. If necessary, translators should be available at the meetings and at availability sessions in order to ensure that all community members have an opportunity to participate, ask questions, and learn about the environmental program or issue.

Stakeholder partnerships can be created to increase community access to decision makers and to ensure that affected community members are included early in the decision-making process. Stakeholder partnerships may include affected community members; federal, state, local, and tribal governments; environmental organizations; nonprofit organizations; academic institutions; and business and industry.

For example, the EPA has established community-based partnerships with other federal agencies and city governments to develop lead abatement projects that involve training community residents to inspect and remediate housing contaminated with lead. The projects train local underemployed or unemployed residents in environmental remediation. The program also serves to empower the community to become involved in restoring their housing stock. (From *The Environmental Justice Strategy: Executive Order 12898*, U.S. EPA, Office of Environmental Justice, EPA/200-R-95-002, April 1995.)

As we've discussed, mailing lists that must be compiled to meet a number of different environmental program requirements should include all stakeholders. Special efforts may be required to include local environmental justice organizations, civic associations, religious institutions, and other stakeholder groups that represent minority and low-income populations.

Education and Training Education and training programs can be approached from two directions. Professional, technical, and business personnel can receive training in environmental justice issues. Such training often focuses on public involvement and environmental health-related topics of particular concern in minority and low-income communities. These training programs can be developed with the assistance of affected stakeholders, environmental organizations, and civic associations in order to address the specific needs of the particular community where the environmental justice issue has arisen.

Training efforts are typically designed to ensure that project managers and other technical personnel understand the concept of environmental justice and can effectively communicate with stakeholders as they take part in an ongoing dialogue with the community. Such training programs attempt to provide an awareness and respect for the unique culture, history, and knowledge of the community. The programs are intended to both educate personnel regarding the history of the environmental justice movement and create an awareness of how community considerations such as human health, employment, and social values are related to one another and to the environmental issue. Training also usually emphasizes the importance of utilizing document formats, language, and meeting styles that are appropriate to the given community.

Technical and training assistance can also be supplied directly to the affected community. Technical assistance and training is often undertaken with the input of local organizations and educational institutions.

For example, the regional EPA, Boston Bar Association, several community-based groups, civil rights organizations, and public health and environmental professionals were involved in the formation of the Massachusetts Environmental Justice Network. This network provides pro bono services to low-income and minority communities on environmental issues. The network offers a comprehensive approach to technical assistance concerning environmental hazards and economic development in traditionally underrepresented communities. (From *Environmental Justice 1994 Annual Report*, U.S. EPA, Office of Environmental Justice, EPA/200-R-95-003, April, 1995.)

Low-income and minority communities may need to receive assistance from federal, state, and local governments, as well as other organizations, in order to develop the technical and analytical expertise necessary to participate in environmentally-based decision making. Assistance methods can include supporting or developing programs based out of community environmental justice organizations, minority colleges and universities, and other nonprofit institutions. Additional programs can be developed in order to conduct health, scientific, technical, and policy-oriented studies and analyses on environmental justice issues.

I N D E X

federal facilities programs, 314–325
NPDES program requirements, 286–289
RCRA program requirements, 252–271
USDA policy, 321
Public notice:
  CERCLA program requirements, 273–283
  emergency planning and notification
    requirements, 293–294
  NCP removal action requirements,
    284–286
  NPDES permitting requirements, 287–288
  RCRA program requirements, 255–271
Public participation, *see also* Community
  relations
  expanded, under RCRA, 254–258
Public perceptions:
  regarding crisis communication, 114
  regarding risk, 102, 104–105
  role of, 6, 30–31, 207–213

Questionnaires, 86, 197–198
Questions and answers, 99, 130–132

Regulations as drivers of need for
  community relations, 10–12
Relationships, importance in community
  relations, 3, 78–79, 120–121
Research methods, *see also* Community
  assessment techniques
  one-on-one interviews, 88
  questionnaires, 86, 197–198
  to evaluate effectiveness of community
    relations program, 196–199
  to supplement communication design,
    85–88
Resource Conservation and Recovery Act
  (RCRA):
  closure and post-closure activities,
    263–266
  corrective action activities, 266–270
  corrective action program, 253
  corrective measures study, 268
  description of, 23
  draft permit stage, 258–259
  driver of need for community relations,
    10–11
  enforcement orders, 270
  environmental justice policy, 270–271,
    327–328
  expanded public participation, 10, 254–258
  final remedy selection, 269
  permit application, 257–258
  permit modification, 259
  permit review, 258

proposed remedy selection, 268
public involvement program requirements,
  255–271
RCRA facility assessment, 258
RCRA facility investigation, 258
Restoration advisory boards, 321–324
Risk, *see also* Risk communication
  perceptions of, 102, 104–105
  control issues, 104–105
Risk-based approach to cleanup, 12, 312–313
Risk communication:
  Allen, Frederick W., 101
  barriers to effective, 103–104
  Covello, Vincent, 101
  definition of, 100–102
  perceptions of, 102, 104–105
  practices and process, 101
  principles of, 101–106
  role of media, 105–106
  "seven cardinal rules of," 101–102
  Slovic, 101
  spokespersons for, 102–103
  von Winterfeldt, 101
Risk Management Program:
  accidental release prevention provisions,
    297–299
  alternative release scenarios, 301
  driver of need for community relations,
    11, 15
  geographic extent of concern, 51
  program requirements, 297–303
  public availability of risk management
    plans, 302–303

Safety-Kleen Corp., 242–245
Sandman, Peter M., 105
SARA Title III, *see* Emergency Planning and
  Community Right-to-know Act
  (EPCRA)
School programs, 157–158
Site-specific advisory boards, 321–324
Small group meetings, 87, 149–150
Social trends as drivers of need for
  community relations, 12–16
Speakers bureau, 157
Spokespersons:
  importance when communicating sensitive
    information, 90–91
  crisis communication, 108–109
  influence on message, 83
  in risk communication, 102–103
Stakeholder(s):
  agendas, 38–39
  controlling, dealing with, 177–178